SpringerBriefs in Molecular Science

SpringerBriefs in Electrical and Magnetic Properties of Atoms, Molecules, and Clusters

Series Editor

George Maroulis, Department of Chemistry, University of Patras, Patras, Greece

More information about this subseries at http://www.springer.com/series/11647

Alexander S. Sharipov · Boris I. Loukhovitski ·
Ekaterina E. Loukhovitskaya

Influence of Internal Degrees of Freedom on Electric and Related Molecular Properties

Springer

Alexander S. Sharipov
Central Institute of Aviation Motors
Moscow, Russia

Boris I. Loukhovitski
Central Institute of Aviation Motors
Moscow, Russia

Ekaterina E. Loukhovitskaya
N.N. Semenov Federal Research Center
for Chemical Physics of Russian Academy
of Science
Moscow, Russia

ISSN 2191-5407   ISSN 2191-5415   (electronic)
SpringerBriefs in Molecular Science
ISSN 2730-7751   ISSN 2730-776X   (electronic)
SpringerBriefs in Electrical and Magnetic Properties of Atoms, Molecules, and Clusters
ISBN 978-3-030-84631-2   ISBN 978-3-030-84632-9   (eBook)
https://doi.org/10.1007/978-3-030-84632-9

This Springer imprint is published by the registered company Springer Nature Switzerland AG
The registered company address is: Gewerbestrasse 11, 6330 Cham, Switzerland

There are a thousand thoughts lying within a man that he does not know till he takes up a pen to write.

—William Makepeace Thackeray
(The History Of Henry Esmond, Esq., 1852)

Preface

Electric properties of molecules that have been extensively studied in the past continue to attract the attention of researchers engaged in various fields of molecular and chemical physics. This is due to the fact that electric properties (mainly, dipole moment and dipole polarizability) determine the variety of optic and electric phenomena in molecular gases, and play a dominant role in intermolecular and electron-molecular interactions. In this regard, it is not surprising that thanks to numerous experimental studies and theoretical efforts, a huge body of knowledge on the electric properties of different molecules, molecular and atomic clusters has been accumulated to date. However, it should be stressed that most of these data relate to molecules in the lowest quantum (rotational, vibrational, and electronic) levels, predominantly populated at room temperature, whereas excitation of internal degrees of freedom of molecules can significantly affect their electric and optic properties.

In this book on the example of a set of about 100 different species, including diatomic, polyatomic molecules and atomic clusters, we discuss the effect of the excitation of rotational, vibrational, and electronic degrees of freedom on the basic electrical properties (dipole moment and static polarizability) and, as a consequence, on molecular optical and transport properties and reactivity. We hope that this brief survey will be a useful contribution to the theory and practice of calculating state-specific electric properties.

The present book mainly summarizes the research experience in the molecular electric response properties gained in recent years in Central Institute of Aviation Motors (Moscow, Russia), but the presentation is surely put in the context of previous studies on related problems. We would like to dedicate this contribution to the memory of Prof. Alexander M. Starik, brilliant scientist, effective group leader, and esteemed colleague, under whose guidance we were lucky to obtain the most of the described results.

Also, we thank our co-workers Alexey V. Pelevkin for numerous fruitful recommendations and assistance with multireference quantum chemistry machinery and Ilya V. Arsentiev for provided results of nonequilibrium oxygen plasma composition behind a shock wave front.

This work was funded by the Russian Foundation for Basic Research (project number 20-38-70014). Chapter 6 was also prepared as part of the implementation of the Program for the creation and development of a world-class scientific center "The Supersonic" (Sverkhzvuk) for 2020–2025 with financial support from the Ministry of Education and Science of Russia (agreement No. 075-15-2021-605 dated June 24, 2021).

Moscow, Russia   Alexander S. Sharipov
June 2021   Boris I. Loukhovitski
                                                                              Ekaterina E. Loukhovitskaya

Contents

1   **Introduction**   1
     References   2

2   **Dependences of Potential Energy and Electric Properties**
     of Molecule on Nuclear Displacements   5
     2.1   Diatomic Molecules   6
         2.1.1   Potential Energy Curves   8
         2.1.2   Dipole Moment Functions   8
         2.1.3   Dipole Polarizability Functions   12
     2.2   Polyatomic Molecules   16
         2.2.1   Sections of DMS and DPS Along the Vibrational
              Coordinates   18
     References   19

3   **Energy Levels and State-Specific Electric Properties**   23
     3.1   Diatomic Molecules   23
         3.1.1   Rovibrational Levels   25
         3.1.2   Dipole Moment   26
         3.1.3   Dipole Polarizability   29
     3.2   Polyatomic Molecules   34
         3.2.1   Vibrational Levels   37
         3.2.2   Dipole Moment   38
         3.2.3   Dipole Polarizability   43
     3.3   Approximation Formulae for the Effect of the Zero-Point
         Vibrations   48
     References   54

4   **Influence of Vibrational and Rotational Degrees of Freedom**
     of Molecules on Their Optical Properties   57
     4.1   Diatomic Molecules   58
     4.2   Polyatomic Molecules   62
     References   65

5   Polarizability of Electronically Excited States 67
     References 73

**6   The Effect of Nonequilibrium in Internal Degrees of Freedom
     of Molecules on Their Physical Properties** 75
     6.1   Refractive Index 75
           6.1.1   Polarizability of a Two-Level System 76
           6.1.2   Influence of Dissociation on Refractivity 77
           6.1.3   Refractivity of a Nonequilibrium Reacting Gas 78
     6.2   Intermolecular Potential 82
     6.3   Rate Constants of Elementary Reactions 86
     6.4   Transport Properties 91
     References 94

7   Conclusions and Future Prospects 97
     References 98

Acronyms

aug-cc-pVTZ	Dunning's correlation consistent triple-zeta basis set with diffuse functions
B2PLYP	Grimme's hybrid density functional with perturbative second-order correlation
B3LYP	Becke's three-parameter hybrid functional
B97-2	Wilson, Bradley and Tozer's modification to B97 functional
CASSCF	Complete active space self-consistent field
CC	Coupled-cluster method
CCSD	Coupled-cluster method with single and double excitations
CCSD(T)	Coupled-cluster method with single and double excitations supplemented with a perturbative correction for triple excitations
CM	Center of mass
DFT	Density functional theory
DMF	Dipole moment function $\mu(r)$
DMS	Dipole moment surface
DP	Dielectric permittivity
DPF	Dipole polarizability function $\alpha(r)$
DPS	Dipole polarizability surface
DRFM	Dipole-reduced formalism method by Paul and Warnatz
EOM	Equation-of-motion
GD	Gladstone–Dale
IP	Ionization potential
LJ	Lennard-Jones potential
MCSCF	Multiconfiguration self-consistent field
MP2	The second-order Møller–Plesset perturbation theory
MP4(SDQ)	The fourth-order Møller–Plesset perturbation theory involving single, double, and quadruple substitutions
MPP	Minimum polarizability principle
PEF	Potential energy function $U(r)$
PES	Potential energy surface
PP	Method of perturbed potential'
RI	Refractive index

Sadlej pVTZ	Triple-zeta basis set developed by Sadlej for the accurate calculations of polarizability
SW	Shock wave
TDDFT	Time-dependent density functional theory
UP	Method of 'unperturbed potential'
UV	Ultraviolet
ZPV	Zero-point vibrations

Chapter 1
Introduction

Wide interest in molecular electric properties, primarily, dipole moment and polarizability, follows essentially from the importance of these quantities in different areas of molecular physics and in related technical applications (viz., linear and nonlinear optical phenomena, theory of infrared and Raman molecular spectra, dielectric properties of gases, intermolecular and electron-molecule interactions, transport properties, kinetic stability and chemical reactivity, rational design of materials with desired properties) [1–14]. For a long time, this was the convincing motivation for extensive experimental and theoretical investigations of the electric properties of different molecules. As a result, a considerable amount of reliable data has been accumulated for molecular dipole moments and polarizabilities [15–19]. In this connection, it should be emphasized that most of these studies were performed, mainly, for the molecules in low vibrational, rotational, and electronic levels, predominantly populated at room temperature. Particularly, in the case of the routine ab initio predictions of electric properties, the molecules in their equilibrium nuclear configurations of the ground electronic state are considered mainly, and the influence of zero-point vibrational motion is often not even taken into account.

However, it is known that excitation of internal degrees of freedom of molecules (vibrational, rotational, and electronic) can significantly affect their electric properties [20–23]. Such excitation can be realized both in thermally equilibrium heating of gas to high temperatures and in essentially nonequilibrium conditions, relevant to the strong shock waves, expanding jets of high-temperature gas, upper and middle atmospheres of different planets, intense chemical reactions, electric discharges or absorption of high-power laser radiation [24–31].

As for the nature of effect of excitation on electric properties, for vibrational (together with rotational) and electronic degrees of freedom it differs fundamentally. In particular, when exciting vibrational and rotational degrees of freedom, this influence is due to the fact that the observable magnitudes of these parameters at given vibrational (rotational) state are specified by the averaging of the electric property over corresponding vibrational (rotational) wave function [21, 32]. Moreover, even

A. S. Sharipov et al., *Influence of Internal Degrees of Freedom on Electric and Related Molecular Properties*, SpringerBriefs in Electrical and Magnetic Properties of Atoms, Molecules, and Clusters, https://doi.org/10.1007/978-3-030-84632-9_1

zero-point vibrations can appreciably contribute to the molecular dipole moment and polarizability [19]. The effect of electronic excitation on electric properties, in turn, manifests itself through the population of a more diffuse, higher molecular orbitals and, as a result, an enhancement in response electric properties of an electronic structure [33, 34].

Thereby, since the electric properties of molecules in the excited states are usually quite different from those in the ground states, for many fundamentals and applications in chemistry and physics, the knowledge of the state-specific electric properties of molecules both in low and in high quantum states is highly desirable [14, 26, 35]. Such data make it possible to predict the effect of nonequilibrium in internal degrees of freedom and of high temperature on various properties of molecular gases (optic, electric, transport, reactivity).

This book addresses the analysis of precisely these questions. The presentation is based primarily on the research experience of the authors in the relevant area. The choice of molecules to illustrate the considered dependences and effects is somewhat biased and reflects the scientific interests of the authors (mainly, combustion and atmospheric chemistry), however, the methods under consideration can no doubt be successfully applied to arbitrary molecular species.

References

1. Buckingham AD, Long DA (1979) Phil Trans R Soc Lond A 293:239
2. Lane NF (1980) Rev Mod Phys 52:29
3. Delone NB, Krainov VP (1988) Fundamentals of nonlinear optics of atomic gases. Wiley, New York
4. Bonin KD, Kresin VV (1997) Electric-dipole polarizabilities of atoms, molecules, and clusters. World Scientific, Singapore
5. Hohm U (2000) J Phys Chem A 104:8418
6. Terenziani F, Painelli A, Soos ZG (2004) J Comput Methods Sci Eng 4:703
7. Lodi L, Tennyson J (2010) J Phys B: At Mol Opt Phys 43:133001
8. Volksen W, Miller RD, Dubois G (2010) Chem Rev 110:56
9. Blair SA, Thakkar AJ (2013) Chem Phys Lett 556:346
10. Karamanis P, Otero N, Pouchan C, Torres JJ, Tiznado W, Avramopoulos A, Papadopoulos MG (2014) J Comput Chem 35:829
11. Xie C, Oganov AR, Dong D, Liu N, Li D, Debela TT (2015) Sci Rep 5:16769
12. Gould T, Bucko T (2016) J Chem Theory Comput 12:3603
13. Loukhovitski BI, Sharipov AS, Starik AM (2016) J Phys B: At Mol Opt Phys 49:125102
14. Sharipov AS, Loukhovitski BI, Starik AM (2016) J Phys B: At Mol Opt Phys 49:125103
15. Nelson RD Jr, Lide DR Jr, Maryott AA (1967) Selected values of electric dipole moments for molecules in the gas phase. Technical Report, National Standard Reference Data Series, National Bureau of Standards 10
16. Lide DR (ed) (2010) CRC handbook of chemistry and physics, 90th edn. CRC Press
17. Hohm U (2013) J Mol Struct 1054–1055:282
18. Hickey AL, Rowley CN (2014) J Phys Chem A 118:3678
19. Sharipov AS, Loukhovitski BI, Starik AM (2017) J Phys B: At Mol Opt Phys 50:165101(19pp)
20. Riley G, Raynes WT, Fowler PW (1979) Mol Phys 38:877
21. Bishop DM (1990) Rev Mod Phys 62:343

22. Urban M, Sadlej AJ (1990) Theor Chim Acta 78:189
23. Helgaker T, Coriani S, Jorgensen P, Kristensen K, Olsen J, Ruud K (2012) Chem Rev 112:543
24. Cvetanovic RJ (1974) Can J Chem 52:1452
25. Letokhov VS (1988) Appl Phys B 46:237
26. Osipov AI, Uvarov AV (1992) Sov Phys Usp 35:903
27. Lukhovitskii BI, Starik AM, Titova NS (2005) Comb Explos Shock Wav 41(4):386
28. Fridman A (2008) Plasma chemistry. Cambridge University Press, Cambridge, UK
29. Azyazov VN (2009) Quantum Electron 39:989
30. Krasnopolsky VA (2011) Planet Space Sci 59:952
31. Colonna G, D'Ammando G, Pietanza LD, Capitelli M (2015) Plasma Phys Control Fusion 57:014009
32. Bishop DM, Pipin J, Silverman JN (1986) Mol Phys 59:165
33. Medved' M, Budzák Š, Pluta T (2015) Theor Chem Acc 134:78
34. Nanda KD, Krylov AI (2016) J Chem Phys 145:204116
35. Callegari A, Theule P, Muenter JS, Tolchenov RN, Zobov NF, Polyansky OL, Tennyson J, Rizzo TR (2002) Science 297:993

Chapter 2
Dependences of Potential Energy and Electric Properties of Molecule on Nuclear Displacements

It is generally recognized that the excitation of vibrational and rotational states of molecules substantially contributes to the molecular electric properties. This arises from the contribution of nuclear motion and the effect of electric field on the vibrational wave function [1–5]. The point being that the distortion of molecular structure with respect to equilibrium geometry leads to an essential change in molecular properties [6–8] (including electric ones [9, 10]), which affects the result of averaging the characteristic of interest over the corresponding wave function. Moreover, even zero-point oscillations can appreciably contribute to the observable dipole moment and polarizability, especially for large molecules.

Most theoretical studies concerning the calculation of the effect of nuclear motion on electric properties address the analysis of the influence of precisely zero-point vibrations [3, 5, 11–32]. In this case, it is sufficient to investigate the potential energy surface (PES) and electric properties of the atomic system often only in the vicinity of its equilibrium nuclear configuration. However, for multiple applications in physical chemistry and chemical physics, it is the state-specific electric properties of molecules in arbitrary quantum states that interest the researchers.

One should recognize that there are very few theoretical works in which the dependences on the vibrational and rotational quantum numbers of the effective dipole moment [10, 33–37] and polarizability [10, 37–44] are comprehensively investigated. The fact is that, in this case, the detailed explorations of corresponding potential energy function (PEF) or PES (with calculating the energy of rovibrational levels and corresponding wave functions) together with the dipole moment functions/surfaces (DMF/DMS) and dipole polarizability functions/surfaces (DPF/DPS) employing accurate quantum chemical methods in a wide range of internuclear spacings r, including the range corresponding to nuclear motion at high vibrational levels, are required. The specifics of such calculations and a review of the results obtained elsewhere for a number of important molecules are discussed in this Chapter.

© The Author(s), under exclusive license to Springer Nature Switzerland AG 2022
A. S. Sharipov et al., *Influence of Internal Degrees of Freedom on Electric and Related Molecular Properties*, SpringerBriefs in Electrical and Magnetic Properties of Atoms, Molecules, and Clusters, https://doi.org/10.1007/978-3-030-84632-9_2

2.1 Diatomic Molecules

For diatomic molecules, the aforementioned problem due to the dimensionality of the system (the only internuclear distance) reduces to finding the corresponding PEF, DMF, and DPF. The determination of each of these functions, in principle, is a particular challenge for existing theoretical methods. It would seem that using the modern methods of quantum chemistry to calculate the potential for a diatomic molecule (abstracting from the response properties) to be no conceptual difficulty, but in reality even this task is far from trivial. It is enough to mention the so-called dissociation catastrophe associated with the need to correctly take into account static correlation at large interatomic separation [45–47]. In fact, to obtain PEF (and, accordingly, the energies of rovibrational levels) for a diatomic molecule with spectroscopic accuracy, apart from correct accounting of the static and dynamic correlations, extraordinary efforts related to allowance for relativistic, spin-orbit, and nonadiabatic effects are required [48–50]. However, fortunately, for determining the state-specific electric properties, high accuracy of the PEF is not critically needed, therefore, in the present book we will not focus on this aspect closely and proceed to consideration of DMF and DPF.

The DMF is commonly used for the calculations of Einstein coefficients and radiative transition probabilities among vibrational levels of diatomic molecules [51–54]. Withal, the DMF, determined over an ample range of r, is also of immense theoretical value because it can unmask the concealed features of the electronic structure of a molecule [55–57]. As regards the dependence of the polarizability on internuclear distance $\alpha(r)$, it enters the theory of low-energy electron-molecular scattering and the models for excitation of molecular vibrations by electron impact [58–60].

We note, in passing, that in most of the previous works, DMF and DPF were studied in the vicinity of the equilibrium internuclear distance r_e: the values of $\mu(r_e)$ and $\alpha(r_e)$ for many diatomic molecules are extensively investigated and tabulated elsewhere [61–66]. However, reliable data on DMF and DPF over a wider range of r is often more limited due to the specificity of the problem and the difficulties of an adequate description of the electronic structure of deformed molecules [47, 67].

Within this Chapter, we restrict ourselves to the analysis of these functions for a number of diatomic molecules significant for combustion, laser, and atmospheric chemistry. The dipole moment function for CO, NO, OH, HCl, HF, and CH molecules were studied computationally based on the ab initio approach in the past [35, 36, 52, 53, 55, 68–85]. However, the use of differing levels of treatment in these theoretical works resulted in meaningful discrepancies in the behavior of predicted DMFs. Additionally, some researches were targeted on empirical or semiempirical approximations and extrapolations of DMF [56, 86], however, the data obtained in this way cannot be regarded as reliable ones in the extended r range. Hence, accurate first-principle calculations of DMFs over the broad range of r are of immense concern. As for DPF, the data available in the literature for widespread diatomic molecules were reported for a rather narrow range of r near r_e [59, 81, 87]. Only for molecular hydrogen the known data on DPF [38, 88, 89]

Table 2.1 The spin multiplicities $(2S + 1)$ and the number of electrons (N) and orbitals (M) into the active space used for the CASSCF calculations in [10] (Reused with permission from Ref. [10]. Copyright 2016 IOP Publishing Ltd)

Molecule	$2S + 1$	N	M
H_2	1	2	6
N_2	1	10	8
O_2	3	10	8
NO	2	11	8
OH	2	7	7
CO	1	10	10
CH	2	5	5
HF	1	8	8
HCl	1	8	8

can be considered complete. More recently, the authors of this Book have made an attempt to conduct the necessary DMF/DPF analysis in the framework of modern multireference quantum chemical machinery for H_2, N_2, O_2, NO, OH, CO, CH, HF, and HCl molecules [10]. It is the description of these results and comparison with the literature data that follow below.

The dependences of potential energy $U^F(r)$ and dipole moment $\mu^F(r)$ were calculated in [10] with the help of complete active space self-consistent field (CASSCF) method [90] and Sadlej triple-zeta pVTZ basis set [91] for three cases: in the absence of external (probe) electric field $(F = 0)$ and when the external electric field is directed along $(F = \parallel)$ and across $(F = \perp)$ molecule axis. Calculation of these dependences in three versions makes it possible to apply the finite-field method [92] for the polarizability computations. Interestingly enough, the Sadlej basis set was intentionally developed for the accurate and computationally efficient calculations of electric properties of relatively large molecules [93]. The dipole moment, in turn, was straightforwardly derived as the expectation value of the dipole operator for the CASSCF electronic wave function. The orthogonal coordinate system was set in such a way that the OZ axis was directed along the molecular axis from the first atom (in a chemical formula) to the second one (in particular, for CO molecule, from carbon atom to oxygen one). Note also that, within the chosen coordinate system, the nondiagonal elements of the polarizability tensor are zero. The spin multiplicities of molecules $(2S + 1$, where S is the electron spin number) and active space dimensions, adopted for the CASSCF calculations of the ground electronic states of considered molecules, are provided in Table 2.1. All first-principle computations were implemented by using Firefly QC program package [94] partially based on the GAMESS(US) source code [95].

2.1.1 Potential Energy Curves

So long as the CASSCF approach does not capture the dynamic electron correlation and, consequently, cannot provide precise dissociation energies for the diatomic molecules at issue, the calculated $U^F(r)$ dependences are to be scaled as follows:

$$U_{scal}^F(r) = \lambda U^F(r_e + (r - r_e)/\sqrt{\lambda}), \lambda = D_e^{ref}/D_e^{calc}. \tag{2.1}$$

Here D_e^{ref} and D_e^{calc} are the magnitudes of the dissociation energy of the molecules reported in the reference database [96] and those calculated in [10], respectively. This transformation of $U^F(r)$ potential enables us to hold the shape of the potential curve in the vicinity of equilibrium point r_e. Interestingly enough, the introduced scaling procedure is rooted in the fact that the $U^F(r)$ dependence of a diatomic molecule is close to a quadratic one at $r \approx r_e$.

Figure 2.1 displays the resultant potential energy curves for H_2, O_2, and CO molecules together with those calculated elsewhere [54, 97–100]. It should be stressed that, in contrast to [10], the top-level calculations [54, 97–100] were performed by means of extrapolation to the complete basis set limit and allowed for the dynamic electron correlation, spin-orbit, and relativistic effects. Nevertheless, we observe adequate agreement between the scaled results of [10] and more accurate ab initio calculations.

2.1.2 Dipole Moment Functions

Let us remark that the choice of the multireference CASSCF method for electric properties is due to a need to obtain the DMF and DPF in a broad range of r. Undoubtedly, density functional theory (DFT) methods proved to be effective for predicting the basic electric properties of molecules near equilibrium r value at low computational expense [65, 101], however, the use of single-reference DFT methods for this purpose is highly questionable at a rather long internuclear distance [10, 102]. This is illustrated in Fig. 2.2, where $\mu_z^0(r)$ dependence for the CO diatomic molecule, calculated with the help of B3LYP [103], B2PLYP [104], and CASSCF techniques and Sadlej basis set, is depicted in comparison with the value of $\mu_z^0(r_e)$ reported in NIST CCCB database [63]. Herein and hereafter the position of equilibrium distance r_e is pointed by a vertical line. As observed, all three methods under consideration reproduce the dependence $\mu_z^0(r)$ near r_e rather consistently. However, the values of dipole moment obtained employing the B3LYP functional at $r \gg r_e$ do not come to zero at all, whereas the dipole moment of a system of two infinitely distant atoms must be equal to zero [57, 86]. The popular double-hybrid functional B2PLYP predicts the μ_z^0 values at extended internuclear distances somewhat better. But the absolute dipole moment value, calculated through the use of the B2PLYP/Sadlej pVTZ level of theory, is about 3 D at $r \sim 2$ Å, whereas, as we see here, different researchers

Fig. 2.1 The $U^0_{scal}(r)$ potential energy curves for H_2, O_2, and CO molecules obtained in [10] (curves) and results of the first-principle calculations reported elsewhere [54, 97–100] (symbols)

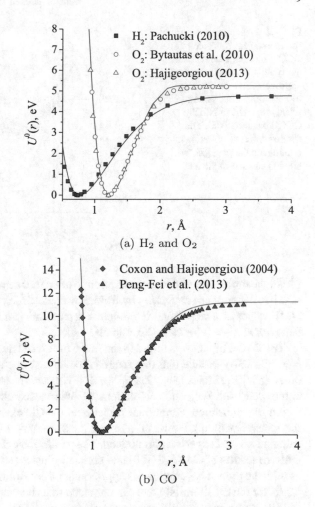

(a) H_2 and O_2

(b) CO

argued that it had to be half that [55, 76, 80, 81]. Meanwhile, the CASSCF method obviously ensures the reasonable dipole moment magnitudes in the wide range of r.

Figures 2.3, 2.4, and 2.5 depict the dependence of dipole moment on the internuclear distance calculated in [10] for considered molecules (OH, CO, NO, CH, HF, and HCl) as well as the theoretical predictions and measurements of other research groups. In general, for all neutral diatomic molecules the following asymptotics hold: $\lim_{r\to 0} \mu^0_z(r) = 0$ (to be more precise, $\mu^0_z(r)$ tends to zero as r^3 [57, 86]) and $\lim_{r\to\infty} \mu^0_z(r) = 0$ [86] (as for a system of two noninteracting atoms). One can infer that the predictions [10] are principally consistent both with the asymptotic $\lim_{r\to\infty} \mu^0_z(r)$ limits and with the findings of other researchers, specifically, near r_e. In particular, for DMF of OH molecule, we see excellent agreement between all theoretical predictions up to the large r spacings. At the same time, based on the calculations [10], we are not able to trace the limiting transition $\lim_{r\to 0} \mu^0_z(r) = 0$

Fig. 2.2 Dependence of μ_z^0 on internuclear distance r for a CO molecule, calculated by applying different theoretical approaches in a broad range of r as well as the value of $\mu_z^0(r_e)$ reported in NIST CCCB database [63]. The data of different researchers obtained in the past [55, 76, 80, 81] are also depicted

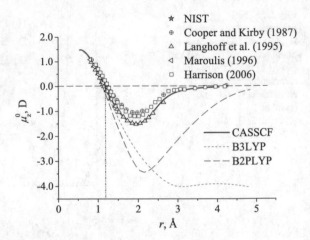

for all the molecules under consideration. Notice that semiempirical approximations of $\mu_z^0(r)$ dependence suggested by Buldakov et al. [56, 86], grounded on the analytic DMF values at a very small internuclear separation, provides the sound asymptotic behavior at $r \to 0$ for OH, CO, and NO molecules.

The DMF for CO was found in the range of internuclear distance $r = 0.54 \div 4$ Å (see Fig. 2.3b). Notice that this range is the widest one among those considered in other ab initio studies [55, 72, 76, 80, 81]. We can observe that the predictions [10] agree well both with the calculations of other researches and with the results of fitting the measured vibrational transitions by Chackerian et al. [74] relevant to the r_e neighborhood. However, at $r > 1.5$ Å a remarkable discrepancy between the calculations and experiment is noticed. The predictions of [10] correlate substantially with the results of CASSCF [80] and coupled-cluster (CC) [81] calculations as well as with the piecewise-continuous approximation by Buldakov and Cherepanov [86].

As for the DMF for NO, we can conclude that the values of μ_z^0 near r_e, predicted in [10], agree well with those reported elsewhere [71, 79, 82] (in this case, $r_e = 1.17$ Å). However for internuclear distance within the range 1.5 Å $< r < 2.5$ Å, the discrepancy between theoretical predictions can be up to one and a half times. For the CH radical, the DMF obtained in [10] is in excellent agreement with earlier electronic structure estimates of Lie et al. [68] and Sun and Freed [77].

As to the DMF for HF, there is a considerable difference between the predictions [35, 52, 73, 84, 85], especially, in the range 1.2 Å $< r < 2.8$ Å (see Fig. 2.5a). Only in the neighborhood of r_e ($r_e = 0.93$ Å) the calculated values of $\mu_z^0(r)$ do correlate well with each other. The fine consistency is observed between the calculations [10] and the results of studies [35, 52, 84, 85].

For the DMF of HCl, one can find superb agreement between the calculations [10] and the DMF data derived from experimental intensity data for vibration-rotational transitions by Ogilvie et al. [73]. However, later calculations [83, 85] and experiment [105] indicate a slightly different $\mu_z^0(r)$ dependence at considerable distances from the equilibrium position. Let us remark that the direct comparison of the cal-

Fig. 2.3 DMF for OH and CO molecules determined in [10] (solid curves) and elsewhere [53, 55, 70, 72, 75, 76, 78, 80, 81] as well as the experiment-based data of Chackerian et al. [74] and reference NIST $\mu_z^0(r_e)$ values [63] (symbols). Semiempirical approximations by Buldakov et al. [56, 86] are shown by dotted curves (Adapted with permission from Ref. [10]. Copyright 2016 IOP Publishing Ltd)

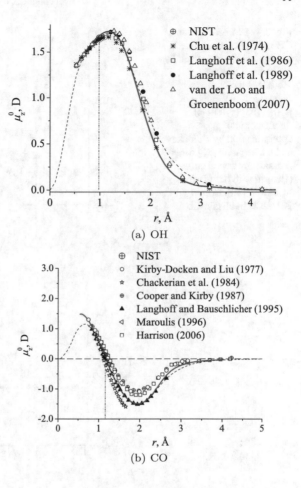

⊕ NIST
✳ Chu et al. (1974)
□ Langhoff et al. (1986)
● Langhoff et al. (1989)
△ van der Loo and Groenenboom (2007)

(a) OH

⊕ NIST
○ Kirby-Docken and Liu (1977)
✶ Chackerian et al. (1984)
⊕ Cooper and Kirby (1987)
▲ Langhoff and Bauschlicher (1995)
◁ Maroulis (1996)
▫ Harrison (2006)

(b) CO

culated $\mu_z^0(r_e)$ values with the reference data on molecular dipole moments [63] is not always justified. The point being that the table data are generally derived from the measurements conducted at a certain temperature and, accordingly, the measured value of dipole moment is to be confronted with the calculated one only upon proper averaging the latter over the nuclear motion.

When comparing the DMF for the molecules in question among themselves, we can conclude that the calculated $\mu_z^0(r)$ dependences have both something in common and some specific features. First, if, for OH, HF, and HCl molecules, the maximal absolute value of $\mu_z^0(r)$ is attained at $r > r_e$, for CO, NO, and CH diatomics, on the contrary, this value is reached at $r < r_e$. Secondly, for CO, NO, and CH molecules, there is an internuclear separation r_s, at which their DMFs change sign. Particularly, for CO and NO molecules, the r_s value is a little larger than r_e. For CH, DMF reverses the sign at $r \approx 1.8$ Å, while r_e equals 1.15 Å. This peculiarity (reversal of DMF sign) is associated with the inherent electronic structure of these molecules [55, 57].

Fig. 2.4 DMF for NO and CH molecules determined in [10] (solid curves) and by other researchers [68, 71, 77, 79, 82] as well as the reference NIST values of $\mu_z^0(r_e)$ [63] (symbols). Semiempirical approximation by Buldakov and Cherepanov [86] for NO is shown by dotted curve (Reused with permission from Ref. [10]. Copyright 2016 IOP Publishing Ltd)

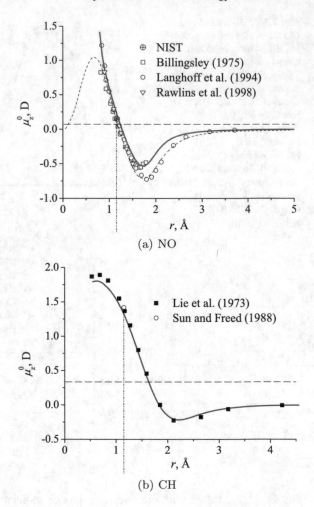

(a) NO

(b) CH

Virtually, the sign change of the DMF with increase in the internuclear separation is a manifestation of the migration of the charge inside a stretching bond (molecule).

2.1.3 Dipole Polarizability Functions

As regards the DPF for molecules at issue, it was computed within the finite-field method, enabling one to obtain static polarizability via the induced dipole moment [92]. Notice that the evidence on the dependence of polarizability on the internuclear distance in diatomic molecule is much scarcer compared to data on the DMF. Only for H_2 molecule the existing data on DPF in the broad range of r [38, 39, 88, 89] can be assumed as conclusive. For N_2 and CO molecules, the polariz-

Fig. 2.5 DMF for HF and HCl calculated in [10] (solid curves) and by other researchers [35, 52, 73, 83–85, 105] as well as the reference NIST values of $\mu_z^0(r_e)$ [63] (symbols)

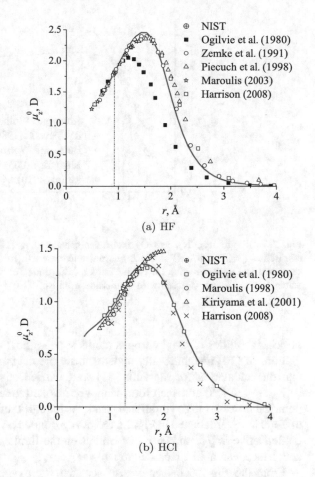

(a) HF

(b) HCl

ability values were obtained only in some neighborhood of r_e [81, 87]. Shown in Figs. 2.6, 2.7, 2.8, and 2.9 are the DPFs predicted in [10] and elsewhere, together with the reference data at equilibrium internuclear distances recommended by Miller in the CRC ("rubber") handbook [106]. For comparison purposes, the asymptotic values of $\alpha(r \to \infty)$, based on the sum of polarizabilities of the noninteracting atoms that make up the molecule [106], are also provided.

Since there are no generally accepted reference data on $\alpha(r_e)$ for OH and CH radicals, their polarizabilities at $r = r_e$, required for the validation of the methodology [10], were estimated in [10] with the help of high-level ab initio CCSD model, which has been proven to be rather accurate in predicting electric properties [64, 65]. Additionally, the value of $\alpha(r_e)$ for OH calculated at the MP4(SDQ)//B97-2/Sadlej pVTZ level of theory was adopted from [107] for comparison.

From the plots shown in Fig. 2.6 one can observe for the H_2 molecule perfect agreement between the predictions [10] and the data of other researchers [38, 39, 88] together with the reference value of $\alpha(r_e)$ and, at $r > 4$ Å and with the sum

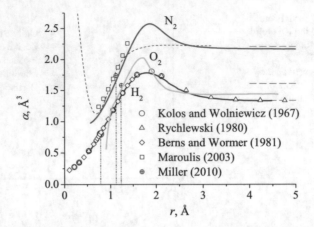

Fig. 2.6 DPF for H_2, N_2, and O_2 molecules calculated in [10] (solid curves) and by other researchers [38, 39, 87, 88] and the relevant reference $\alpha(r_e)$ data [106] (symbols) as well as semiempirical approximation for N_2 by Temkin [59] (dotted curve) and asymptotic $\alpha(r \to \infty)$ values (dashed lines) (Reused with permission from Ref. [10]. Copyright 2016 IOP Publishing Ltd)

of polarizabilities of two hydrogen atoms ($\alpha(r \to \infty)$) [106]. But for N_2, the predictions of [10] insignificantly underestimate the data of Maroulis [87]. Note also that the calculations of Maroulis [87] were carried out only in the vicinity of r_e ($0.7 < r < 1.4$ Å) and, therefore, cannot be invoked for extracting the state-specific polarizabilities for high vibrational states. As for early semiempirical approximation of Temkin [59], depicted in Fig. 2.6 too, it is grounded on the known values of α and its derivative $\partial\alpha/\partial r$ at $r = r_e$, as well as on the limits $\alpha(r \to 0)$ and $\alpha(r \to \infty)$, and, hence, can not be regarded reliable at $r > r_e$.

From the Fig. 2.6 one may also see that for oxygen molecule, the value of $\alpha(r \to \infty)$, predicted in [10], is lower than the sum of polarizabilities of two ground-state $O(^3P)$ atoms by a factor of 1.11. Nevertheless, the implemented methodology reproduces with tolerable accuracy the reference $\alpha(r_e)$ value, reported by Miller in the CRC handbook [106]. A comparable situation is observed for the DPF of NO (see Fig. 2.7). As for the polarizability of CO, an acceptable coincidence of calculations [10] with the $\alpha(r \to \infty)$ value based on the reference data [106] as well as with the accurate data of Maroulis [81] occurs. Note also that the DPFs predicted in [10] for CH, OH, HCl, and HF molecules, coincide quite well with calculations [83, 84], asymptotic $\alpha(r \to \infty)$ limits, and with $\alpha(r_e)$ values reported in [106] or estimated elsewhere.

From the plots shown in Figs. 2.6, 2.7, 2.8, and 2.9 it follows that the general behavior of DPF is almost similar for all molecules in question. At values of r close to r_e, polarizability drastically rises with the internuclear distance approaching its maximum at $r_{max} \approx 1.9\, r_e$ for H-containing molecules and at $r_{max} \approx 1.5\, r_e$ for N_2 and O_2. At $r > r_{max}$, polarizability decreases, reaching its asymptotic value that equals the sum of the polarizabilities of isolated atoms included in the molecule.

Fig. 2.7 DPF for CO and NO molecules calculated in [10] (solid curves) and by Maroulis [81] as well as the relevant reference $\alpha(r_e)$ data [106] (symbols) and asymptotic $\alpha(r \to \infty)$ values (dashed lines) (Reused with permission from Ref. [10]. Copyright 2016 IOP Publishing Ltd)

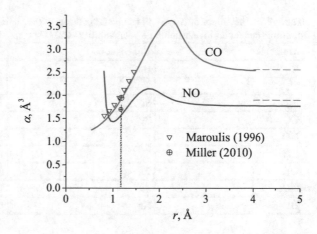

Fig. 2.8 DPF for OH and CH molecules predicted in [10] and asymptotic $\alpha(r \to \infty)$ values (dashed lines) together with the $\alpha(r_e)$ values calculated in [10, 107] (symbols)

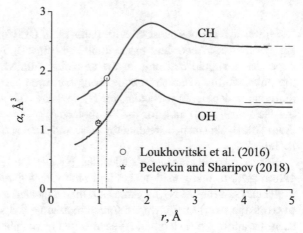

Fig. 2.9 DPF for HF and HCl molecules predicted in [10] (solid curves) and by other researchers [83, 84] together with the reference data for $\alpha(r_e)$ [106] (symbols) and asymptotic $\alpha(r \to \infty)$ values (dashed lines)

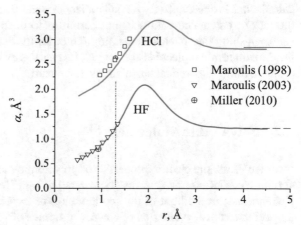

Table 2.2 The values of $\alpha(r \to 0)$ estimated in terms of the "united atom limit" approach for the diatomic molecules in question. The polarizability values for "united atoms" are adopted from [106]

Molecule	United atom	$\alpha(r \to 0)$
H_2	He	0.205
O_2	S	2.90
N_2	Si	5.38
CO	Si	5.38
NO	P	3.63
CH	N	1.10
OH	F	0.557
HCl	Ar	1.641
HF	Ne	0.396

However, a curious question is the behavior of DPF for diatomics at r substantially smaller than r_e. According to the conjecture [9, 58, 59, 108], at $r \to 0$, the $\alpha(r)$ dependence should converge to the so-called "united atom limit" conforming to the polarizability of an atom comprising the same number of protons as the given diatomic molecule. In particular, for H_2 we have $\alpha(r \to 0) = 0.205$ Å^3 (as that for the He atom [106]) and, for the O_2 molecule, $\alpha(r \to 0) = 2.90$ Å^3 (as that for S atom [106]). The respective data for these and the rest diatomic molecules are listed in Table 2.2.

For a number of considered molecules (H_2, CH, OH, HCl, HF), the calculated $\alpha(r)$ values seem to really tend at small r to the values given in Table 2.2. However, for molecules such as O_2, N_2, CO, and NO (their expected $\alpha(r \to 0)$) values significantly override the $\alpha(r_e)$ ones), we are forced to assume a sharp increase in polarizability as atoms approach each other. As seen, a hint of such behavior in quantum chemical calculations is observed only for NO molecule, whereas for O_2, N_2, and CO molecule the lack of calculated data does not permit us to trace such a passage to the limit for a vanishing internuclear separation. Thus, for the definitive conclusion about the DPF for diatomic molecules at $r \ll r_e$, it is necessary to significantly expand the r range in future theoretical calculations downward.

2.2   Polyatomic Molecules

Over the years, the electric properties of polyatomic molecules [106, 109–112] and certain types of atomic (covalent) clusters [113–117] were extensively measured. Nevertheless, in the light of the shortcomings of existing experimental techniques, a priori theoretical evaluation of electric attributes of various polyatomic species is still in the focus of research groups engaged in the field of computational quantum chemistry. Particularly, along with semiempirical and empirical estimates [66, 118,

119], much attention has been given to the accurate first-principles prediction of the electric properties of polyatomic molecules [64, 65, 120–122], molecular [123–125] and atomic [126–129] clusters. Such calculations are usually performed for the molecules, atomic and molecular clusters in their equilibrium nuclear geometry, and the influence of nuclear motion is neglected as a rule.

The fact is that the accurate estimating of the influence of the vibrational motion of nuclei on dipole moment and polarizability for polyatomic molecules is more difficult and resource-intensive than for diatomics, so long as this demands, in the general case, the accurate exploration of corresponding PES, DMS, and DPS [130–135]. Frequently, instead of determining the state-specific electric properties of polyatomic systems only the effect of zero-point vibrational (ZPV) motion is examined or coarse estimates are used [5, 15, 18, 21, 22, 26, 28]. However, due to the need for numerous applications to know the state-specific electric properties of molecules in excited vibrational levels [33, 136–138], the development of a simplified methodology that makes it possible to estimate these characteristics for polyatomic structures based on conventional quantum chemical calculations rather accurately at the moderate computational expense, at least, for not very high vibrational states, would be strongly desirable.

Consider a variant of such a methodology proposed by us recently in [37]. To find the contribution of molecular vibrations to electric properties, the PES, DMS, and DPS must be meticulously explored in the vicinity of the energy minimum point corresponding to the equilibrium geometry of a polyatomic molecule. To this end, the calculations with the B97-2 functional [139] were conducted. As evinced in [139], this DFT functional allows one to predict the electric properties with an accuracy close to that of Brueckner Doubles method [140], assessed as the reference one to predict these characteristics. This is not surprising since the B97-n family of functionals, in contrast to numerous later DFT developments excessively focused on high-precision thermochemistry, can ensure a physically sound spatial distribution of electron density in finite-size atomic systems [101, 141, 142]. The B97-2 functional was also effectively applied in the past for the polarizability calculations for different molecules [46, 101, 143] and atomic clusters [129] as well for the analysis of the influence of vibrational anharmonicity on the thermodynamic functions of polyatomic molecules [144, 145].

The Sadlej pVTZ triple-zeta basis set [91], specially designed for the accurate polarizability calculations [93], can be used for polyatomic molecules as well as for diatomic ones. For molecules containing elements for which the original Sadlej pVTZ basis set is not available, the diffuse-augmented Dunning's correlation consistent triple-zeta basis set (aug-cc-pVTZ) can be applied. All computations of [37] were implemented by using Firefly QC program package [94] partially based on the GAMESS(US) source code [95].

The methodology, applied in [37] for determining the dipole moment μ and static polarizability α values of polyatomic molecule in given vibrational state with vibrational quantum numbers $V_1, ... V_N$ (corresponding to vibrational modes with frequencies $\nu_1, ... \nu_N$, where N is a number of vibrational modes in molecule), was grounded on the approach applied for diatomic molecules (see [10]). The manifold of vibrational states of a polyatomic molecule was represented by the set of separable one-dimensional oscillators, specifying the isolated modes. For this purpose, the spatial structures of all considered compounds were optimized with tight convergence criteria at the UB97-2/Sadlej pVTZ level of theory. Normal mode analysis was done at the same level of treatment within the harmonic approximation.

For each mode m, the dependences of potential energy U_m^F and dipole moment μ_m^F on the normal vibrational coordinate r were determined with the same theoretical model for four cases: the external electric field is absent ($F = 0$) and when the external electric field is directed along one of the coordinate axes ($F = X, Y, Z$). Calculation of these dependences in three variants enables one to apply the finite-field [92] method for the polarizability computations. By this, the values of μ_m^F were calculated as the expectation values of the dipole operator. Displacement vectors of normal vibrations for each atom (nucleus) and the reduced masses M_m were also obtained for each mode. In so doing, it was supposed that the atoms in a molecule move rectilinearly along the displacement vectors of normal vibrations. Of course, this supposition is approximate at high vibrational energy. However, if one considers only low-lying levels (within the preselected energy cutoff criterion E_{cut}), this "linear" assumption permits one to bypass both the detailed examination of corresponding PES and the need to define the ad-hoc set of curvilinear coordinates for each specific system in question (see [12, 34, 146] for more information about using curvilinear coordinates in such problems).

2.2.1 Sections of DMS and DPS Along the Vibrational Coordinates

Let us demonstrate now by the example of CO_2 triatomic molecule that the DFT approach is suitable for finding the DMS and DPS of polyatomic molecules described above. To this end, the calculations of some DMS and DPS sections for CO_2 performed at the B97-2/Sadlej pVTZ computational level were compared with the highly accurate ab initio data. Figure 2.10 displays the increments in dipole moment $\Delta\mu$ and polarizability $\Delta\alpha$ of CO_2 during the motion, corresponding to ν_3 mode (asymmetric stretch), against the parameter of asymmetric displacement ΔR. As seen, the DFT calculations [37] for the increments of dipole moment $\Delta\mu = \mu(\Delta R) - \mu_e$ and polarizability $\Delta\alpha = \alpha(\Delta R) - \alpha_e$ agree nicely with the highly-correlated MP2 and CCSD(T) results [147]. The estimates of the $\Delta\alpha$ increment at the MP4(SDQ)/Sadlej pVTZ level of treatment, performed in [37], are also given in Fig. 2.10b for comparison. They were computed by means of the finite-field approach via the induced dipole moment [92]. Thus, based on the example of an asymmetric stretching mode of CO_2 molecule, we can infer that the proposed DFT methodology (B97-2/Sadlej pVTZ computational level) allows us to reach the accuracy of the DMS and DPS determination akin to that achieved when using more accurate ab initio techniques. In the next Chapter, this will also be demonstrated for the position of energy levels of the bending modes of polyatomic molecules.

Fig. 2.10 The increment of dipole moment $\Delta\mu$ (**a**) and polarizability $\Delta\alpha$ (**b**) for the asymmetric stretch of CO_2 molecule against the asymmetric displacement parameter ΔR: calculations at the UB97-2/Sadlej pVTZ theoretical model (solid curves) and calculations at high levels of treatment by Haskopoulos and Maroulis [147] (MP2 and CCSD(T)) and of [37] (MP4(SDQ)) (symbols) (Adapted with permission from Ref. [37]. Copyright 2017 IOP Publishing Ltd)

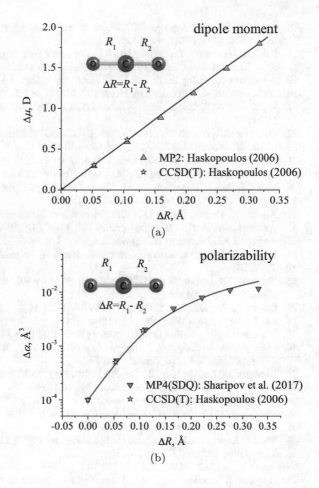

(a)

(b)

References

1. Riley G, Raynes WT, Fowler PW (1979) Mol Phys 38:877
2. Bishop DM, Pipin J, Silverman JN (1986) Mol Phys 59:165
3. Bishop DM (1990) Rev Mod Phys 62:343
4. Chou CC, Jin BY (2009) Theor Chem Acc 122:313
5. Egidi F, Giovannini T, Piccardo M, Bloino J, Cappelli C, Barone V (2014) J Chem Theory Comput 10:2456
6. Wodtke AM (2001) Phys Chem Earth Part C 26:467
7. Krisyuk BE (2004) J Mol Struct (THEOCHEM) 677:77
8. Chirila CC, Lein M (2006) Phys Rev A 74:051401
9. Buldakov MA, Cherepanov VN (2004) J Comput Methods Sci Eng 4:237
10. Loukhovitski BI, Sharipov AS, Starik AM (2016) J Phys B: At Mol Opt Phys 49:125102
11. Adamowicz L (1988) J Chem Phys 89:6305
12. Russell AJ, Spackman MA (1997) Mol Phys 90:251
13. Christiansen O, Hattig C, Gauss J (1998) J Chem Phys 109:4745
14. Bishop DM, Norman P (1999) J Chem Phys 111:3042

15. Ingamells VE, Papadopoulos MG, Raptis SG (1999) Chem Phys Lett 307:484
16. Maroulis G (2000) Chem Phys Lett 318:181
17. Ruud K, Jonsson D, Taylor PR (2000) Phys Chem Chem Phys 2:2161
18. Ruud K, Astrand PO, Taylor PR (2000) J Chem Phys 112:2668
19. Avramopoulos A, Ingamells VE, Papadopoulos MG, Sadlej AJ (2001) J Chem Phys 114:198
20. Avramopoulos A, Papadopoulos MG (2002) Mol Phys 100:821
21. Jug K, Chiodo S, Calaminici P, Avramopoulos A, Papadopoulos MG (2003) J Phys Chem A 107:4172
22. Pederson MR, Baruah T, Allen PB, Schmidt C (2005) J Chem Theory Comput 1:590
23. Costa MF, Ribeiro MCC (2006) Quim Nova 29:1266
24. Xenides D, Maroulis G (2006) J Phys B: At Mol Opt Phys 39:3629
25. Santiago E, Castro MA, Fonseca TL, Mukherjee PK (2008) J Chem Phys 128:064310
26. Luis JM, Reis H, Papadopoulos M, Kirtman B (2009) J Chem Phys 131:034116
27. Maroulis G (2011) Int J Quantum Chem 111:807
28. Aguado A, Vega A, Balbas LC (2011) Phys Rev B 84:165450
29. Avramopoulos A, Reis H, Papadopoulos MG, Conf AIP (2012) Proceedings 1504:616
30. Arapiraca AF, Mohallem JR (2014) Chem Phys Lett 609:123
31. Deiß M, Drews B, Denschlag JH, Bouloufa-Maafa N, Vexiau R, Dulieu O (2015) New J Phys 17:065019
32. Sharipov AS, Loukhovitski BI, Pelevkin AV, Kobtsev VD, Kozlov DN (2019) J Phys B: At Mol Opt Phys 52:045101
33. Osipov AI, Panchenko VY, Filippov AA (1984) Sov J Quant Electron 14:1259
34. Fowler PW, Raynes WT (1981) Mol Phys 43:65
35. Piecuch P, Spirko V, Kondo AE, Paldus J (1998) Mol Phys 94:55
36. Li G, Gordon IE, Le Roy RJ, Hajigeorgiou PG, Coxon JA, Bernath PF, Rothman LS (2013) J Quant Spectrosc Radiat Transfer 121:78
37. Sharipov AS, Loukhovitski BI, Starik AM (2017) J Phys B: At Mol Opt Phys 50:165101(19pp)
38. Kolos W, Wolniewicz L (1967) J Chem Phys 46:1426
39. Rychlewski J (1980) Mol Phys 41:833
40. Hunt JL, Poll JD, Wolniewicz L (1984) Can J Phys 62:1719
41. Maroulis G, Bishop D (1986) Mol Phys 58:273
42. Marti J, Andres JL, Bertran J, Duran M (1993) Mol Phys 80:625
43. Maroulis G, Makris C (1997) Mol Phys 91:333
44. Tang LY, Yan ZC, Shi TY, Babb JF (2014) Phys Rev A 90:012524
45. Helgaker T, Coriani S, Jorgensen P, Kristensen K, Olsen J, Ruud K (2012) Chem Rev 112:543
46. Cohen AJ, Mori-Sanchez P, Yang W (2012) Chem Rev 112:289
47. Mayer I (2013) Simple theorems, proofs, and derivations in quantum chemistry. Springer Science & Business Media
48. Bytautas L, Ruedenberg K (2010) J Chem Phys 132:074109
49. Szalay PG, Holka F, Fremont J, Rey M, Peterson KA, Tyuterev VG (2011) Phys Chem Chem Phys 13:3654
50. Koput J (2015) J Comput Chem 36:2219
51. Hilborn RC (1982) Am J Phys 50:982
52. Zemke WT, Stwalley WC, Langhoff SR, Valderrama GL, Berry MJ (1991) J Chem Phys 95:7846
53. van der Loo MPJ, Groenenboom GC (2007) J Chem Phys 126:114314
54. Peng-Fei L, Lei Y, Zhong-Yuan Y, Yu-Feng G, Tao G (2013) Commun Theor Phys 59:193–198
55. Harrison JF (2006) J Phys Chem A 110:10848
56. Buldakov MA, Kalugina YN, Koryukina EV, Cherepanov VN (2007) Atmos Ocean Opt 20:15
57. Buldakov MA, Cherepanov VN, Koryukina EV, Kalugina YN (2009) J Phys B: At Mol Opt Phys 42:105102 (5pp)
58. Chandra N, Temkin A (1976) Phys Rev A 13:188
59. Temkin A (1978) Phys Rev A 17:1232
60. Itikawa Y (1997) Int Rev Phys Chem 16:155

61. Nelson RD Jr, Lide DR Jr, Maryott AA (1967) Selected values of electric dipole moments for molecules in the gas phase. Technical report, National Standard Reference Data Series, National Bureau of Standards 10
62. Lide DR (ed) (2010) CRC Handbook of chemistry and physics, 90th edn. CRC Press
63. Johnson RD III, NIST (2010) Computational chemistry comparison and benchmark database, NIST standard reference database number 101 release, 15a edn.
64. Hickey AL, Rowley CN (2014) J Phys Chem A 118:3678
65. Wu T, Kalugina YN, Thakkar AJ (2015) Chem Phys Lett 635:257
66. Pereira F, de Souse JA (2018) J Cheminform 10:43
67. Kedziora GS, Barr SA, Berry R, Moller JC, Breitzman TD (2016) Theor Chem Acc 135:79
68. Lie GC, Hinze J, Liu B (1973) J Chem Phys 59:1887
69. Sileo RN, Cool T (1976) J Chem Phys 65:117
70. Chu S, Yoshimine M, Liu B (1974) J Chem Phys 61:5389
71. Billingsley FP II (1975) J Chem Phys 62:864
72. Kirby-Docken K, Liu B (1977) J Chem Phys 66:4309
73. Ogilvie JF, Rodwell WR, Tipping RH (1980) J Chem Phys 73:5221
74. Chackerian C Jr, Farrenq R, Guelachvili G, Rossetti C, Urban W (1984) Can J Phys 62:1579
75. Langhoff SR, Werner HJ, Rosmus P (1986) J Mol Spectrosc 118:507
76. Cooper D, Kirby K (1987) J Chem Phys 87:424
77. Sun H, Freed KF (1988) J Chem Phys 88:2659
78. Langhoff SR, Bauschlicher CW Jr, Taylor JPR (1989) J Chem Phys 91:5953
79. Langhoff SR, Bauschlicher CW Jr, Partridge H (1994) Chem Phys Lett 223:416
80. Langhoff S, Bauschlicher C Jr (1995) J Chem Phys 102:5220
81. Maroulis G (1996) J Phys Chem 100:13466
82. Rawlins WT, Person JC, Fraser ME, Miller SM, Blumberg WAM (1998) J Chem Phys 109:3409
83. Maroulis G (1998) J Chem Phys 108:5432
84. Maroulis G (2003) J Mol Struct (THEOCHEM) 633:177
85. Harrison JF (2008) J Chem Phys 128:114320
86. Buldakov MA, Cherepanov VN (2004) J Phys B: At Mol Opt Phys 37:3973
87. Maroulis G (2003) J Chem Phys 118:2673
88. Berns RM, Wormer PES (1981) Mol Phys 44:1215
89. Lopez X, Piris M, Nakano M, Champagne B (2014) J Phys B: At Mol Opt Phys 47:015101
90. Roos BO (1987) Adv Chem Phys 68:399
91. Sadlej pVTZ EMSL basis set exchange library. https://bse.pnl.gov/bse/portal
92. Kurtz HA, Stewart JJ, Dieter KM (1990) J Comput Chem 11:82
93. Sadlej AJ (1988) Collec Czech Chem Commun 53:1995
94. Granovsky AA Firefly V. 8.2.0. http://classic.chem.msu.su/gran/firefly/index.html
95. Schmidt MW, Baldridge KK, Boatz JA, Elbert ST, Gordon MS, Jensen JH, Koseki S, Matsunaga N, Nguyen KA, Su S, Windus TL, Dupuis M, Montgomery JA (1993) J Comput Chem 14:1347
96. Burcat A, Ruscic B (2011) Third millennium ideal gas and condensed phase thermochemical database for combustion with updates from active thermochemical tables, ANL-05/20 and TAE 960 Technion-IIT, Aerospace Engineering, and Argonne National Laboratory. Chemistry Division 2005
97. Coxon JA, Hajigeorgiou PG (2004) J Chem Phys 121:2992
98. Bytautas L, Matsunaga N, Ruedenberg K (2010) J Chem Phys 132:074307 (15 pp)
99. Pachucki K (2010) Phys Rev A 82:032509
100. Hajigeorgiou PG (2013) J Chem Phys 138:014309
101. Hait D, Head-Gordon M (2018) J Chem Theory Comput 14:1969
102. Shedge SV, Joshi SP, Pal S (2012) Theor Chem Acc 131:1094
103. Becke AD (1993) J Chem Phys 98:5648
104. Grimme S (2006) J Chem Phys 124:034108 (16 pp)
105. Kiriyama F, Rao BS, Nangia VK (2001) J Quant Spectrosc Radiat Transf 69:35

106. Miller TM (2010) CRC handbook of chemistry and physics, 90th edn., vol. 10, chap. Atomic and molecular polarizabilities. CRC Press, Boca Raton, Florida, pp 193–202
107. Pelevkin AV, Sharipov AS (2018) J Phys D Appl Phys 51:184003
108. Buckingham AD, Watts RS (1973) Mol Phys 26:7
109. Maroulis G, Hohm U (2007) Phys Rev A 76:032504
110. Hackermuller L, Hornberger K, Gerlich S, Gring M, Ulbricht H, Arndt M (2007) Appl Phys B 89:469
111. Lide DR, Haynes WM (eds) CRC handbook of chemistry and physics, 90th edn., vol. 9, chap. Dipole moments. CRC Press, Boca Raton, Florida, pp 50–58
112. Hohm U (2013) J Mol Struct 1054–1055:282
113. de Heer WA (1993) Rev Mod Phys 65:611
114. Benichou E, Antoine R, Rayane D, Vezin B, Dalby FW, Dugourd P, Broyer M, Ristori C, Chandezon F, Huber BA, Rocco JC, Blundell SA, Guet C (1999) Phys Rev A 59:R1
115. Hohm U (2000) Vacuum 58:117
116. Tikhonov G, Kasperovich V, Wong K, Kresin VV (2001) Phys Rev A 64:063202
117. Heiles S, Schäfer R (2014) Dielectric properties of isolated clusters: beam deflection studies. Springer, Netherlands, Dordrecht
118. Malykhanov YB (1995) J Appl Spectrosc 62:83
119. Guha S, Menendez J, Page JB, Adams GB (1996) Phys Rev B 53:13106
120. Paidarova I, Sauer SPA, Conf AIP (2012) Proceedings 1504:695
121. Karne AS, Vaval N, Pal S, Vasquez-Perez JM, Koster AM, Calaminici P (2015) Chem Phys Lett 635:168
122. Kalugina YN, Thakkar AJ (2015) Mol Phys 113:2939
123. Gregory JK, Clary DC, Liu K, Brown MG, Saykally RJ (1997) Science 275:814
124. Maroulis G (2012) Int J Quantum Chem 112:2231
125. Kalugina YN, Sunchugashev DA, Cherepanov VN (2018) Chem Phys Lett 692:184
126. Reis H, Papadopoulos MG, Boustani I (2000) Int J Quantum Chem 78:131
127. Xenides D, Maroulis G (2007) J Comput Methods Sci Eng 7:431
128. Karamanis P, Pouchan C, Maroulis G (2008) Phys Rev A 77:013201
129. Sharipov AS, Loukhovitski BI (2019) Struct Chem 30:2057
130. Bowman JM, Irle S, Morokuma K, Wodtke A (2001) J Chem Phys 114:7923
131. Lodi L, Tolchenov RN, Tennyson J, Lynas-Gray AE, Shirin SV, Zobov NF, Polyansky OL, Csaszar AG, van Stralen JNP, Visscher L (2008) J Chem Phys 128:04430
132. Yurchenko SN (2013) Chem Modell 10:183
133. Delahaye T, Nikitin AV, Rey M, Szalay PG, Tyuterev VG (2015) Chem Phys Lett 639:275
134. Azzam AAA, Lodi L, Yurchenko SN, Tennyson J (2015) J Quant Spectrosc Radiat Transfer 161:41
135. Farasat M, Reza Shojaei SH, Maqsood Golzan M, Farhadi K (2016) J Mol Struct (Theochem) 1108:341
136. Osipov AI, Uvarov AV (1992) Sov Phys Usp 35:903
137. Callegari A, Theule P, Muenter JS, Tolchenov RN, Zobov NF, Polyansky OL, Tennyson J, Rizzo TR (2002) Science 297:993
138. Sharipov AS, Loukhovitski BI, Starik AM (2016) J Phys B: At Mol Opt Phys 49:125103
139. Wilson PJ, Bradley TJ, Tozer DJ (2001) J Chem Phys 115:9233
140. Handy NC, Pople JA, Head-Gordon M, Raghavachari K, Trucks GW (1989) Chem Phys Lett 164:185
141. Medvedev MG, Bushmarinov IS, Sun J, Perdew JP, Lyssenko KA (2017) Science 355:aah5975
142. Marjewski AA, Medvedev MG, Gerasimov IS, Panova MV, Perdew JP, Lyssenko KA, Dmitrienko AO (2018) Mendeleev Commun 28:225
143. Verma P, Truhlar DG (2017) Phys Chem Chem Phys 19:12898
144. Boese AD, Klopper W, Martin JML (2005) Mol Phys 103:863
145. Loukhovitski BI, Sharipov AS, Starik AM (2016) Eur Phys J D 70:250
146. Raynes WT (1988) Mol Phys 63:719
147. Haskopoulos A, Maroulis G (2006) Chem Phys Lett 417:235

Chapter 3
Energy Levels and State-Specific Electric Properties

To obtain the state-specific electric properties of molecules in arbitrary quantum states, it is necessary to average them over the corresponding eigen wave functions. The PEFs/PESs, DMFs/DMSs, and DPFs/DPSs described in the previous Chapter for a number of important molecules will be used for this purpose. Features of such calculations for both diatomic and polyatomic species and an overview of the results obtained are discussed in the present Chapter in the context of the knowledge on the state-specific electric properties of molecules accumulated to date.

3.1 Diatomic Molecules

To find the dipole moment and static polarizability, averaged over the vibrational and rotational motion of a diatomic molecule, the radial vibrational wave functions $\chi^F(r)$ are to be found within the Born-Oppenheimer framework by solving the steady-state nuclear-motion Schrödinger equation for the scaled potential $U_{\text{scal}}^F(r)$ given by Eq. (2.1). This can be expressed as follows [1]:

$$\frac{d^2\chi^F}{dr^2} + \left[\frac{2m}{\hbar^2}(E - U_{\text{scal}}^F(r)) - \frac{J(J+1)}{r^2}\right]\chi^F = 0, \quad (3.1)$$

where m is the reduced mass of atoms in a molecule, $\hbar$ is the reduced Planck constant, E is the energy, and J is the rotational quantum number.

Equation (3.1) can be numerically solved for three cases mentioned previously ($F = 0, \parallel$, and $\perp$, see Sect. 2.1) by using the integration technique suggested initially by Numerov [2–4]. Accordingly, eigen energy values $E_{V,J}^F$ and corresponding wave functions $\chi_{V,J}^F$ for given vibrational V and rotational J quantum numbers can be found.

© The Author(s), under exclusive license to Springer Nature Switzerland AG 2022   23
A. S. Sharipov et al., *Influence of Internal Degrees of Freedom on Electric and Related Molecular Properties*, SpringerBriefs in Electrical and Magnetic Properties of Atoms, Molecules, and Clusters, https://doi.org/10.1007/978-3-030-84632-9_3

Average values of internuclear distance r^F and dipole moment μ^F for the given vibrational-rotational (V, J) state can be determined via the Dirac bra and ket formalism as

$$p_{V,J}^F = \langle \chi_{V,J}^F \mid p^F \mid \chi_{V,J}^F \rangle, \quad p^F = r^F \text{ or } \mu^F. \tag{3.2}$$

The state-resolved polarizabilities, in turn, can be calculated by means of two different approaches rooted in the finite-field method [5, 6]. The first one presumes that the influence of the external (probe) electric field on the interatomic potential is ignorable. Under this approach denoted herein and hereafter as the method of 'unperturbed potential' (UP) the averaged value of polarizability $\alpha_{V,J}^{UP}$ for the specific vibrational-rotational level is stated as follows:

$$\alpha_{V,J}^{UP} = \langle \chi_{V,J}^0 \mid \alpha^{UP}(r) \mid \chi_{V,J}^0 \rangle, \tag{3.3}$$

$$\alpha^{UP}(r) = \left(2\frac{\mu_x^\perp(r) - \mu_x^0(r)}{\mathcal{E}} + \frac{\mu_z^\parallel(r) - \mu_z^0(r)}{\mathcal{E}} \right) / 3. \tag{3.4}$$

Here $\mu_x^\perp(r)$ and $\mu_z^\parallel(r)$ are the projections of dipole moment onto the OX and the OZ axes when the external electric field of strength $\mathcal{E}$ is applied across and along the molecule axis, respectively (the value of $\mathcal{E}$, to ensure the stability of numerical differentiation, should not be too small or too high, it can be set equal to, say, 10^{-3} a.u. [7, 8]), $\mu_x^0(r)$ and $\mu_z^0(r)$ are the projections of dipole moment in the absence of external electric field onto the OX and OZ axes. The linearity of $\mu_i^F(\mathcal{E})$ $(i = x, y, z)$ dependences together with the unwanted appearance of significant nondiagonal elements of polarizability tensor must be checked in the course of ab initio calculations.

The second method of polarizability calculation, introduced earlier in [9, 10], accounts for the perturbation of potential by probe field and is indicated hereafter as 'perturbed potential' (PP) one. It allows one to implicitly treat the impact of the external field on the vibrational wave function. In this case, the state-specific polarizability $\alpha_{V,J}^{PP}$ is determined as

$$\alpha_{V,J}^{PP} = \left(2\frac{\mu_{x,V,J}^\perp - \mu_{x,V,J}^0}{\mathcal{E}} + \frac{\mu_{z,V,J}^\parallel - \mu_{z,V,J}^0}{\mathcal{E}} \right) / 3. \tag{3.5}$$

Here $\mu_{x,V,J}^F$ and $\mu_{z,V,J}^F$ are the projections of dipole moment onto the OX and OZ axes averaged over the vibrational wave function, corresponding to the specific vibrational-rotational (V, J) state and obtained for the potential that is perturbed by an external field exerted along the respective directions. These projections are defined by

$$\mu_{i,V,J}^F = \langle \chi_{V,J}^F \mid \mu_i^F \mid \chi_{V,J}^F \rangle, i = x, z \tag{3.6}$$

It should be underlined that the PP method is relevant to the static electric field, whereas for the alternating electric field, the effect of perturbation of vibrational wave function by an external field goes to zero with the increase of field oscillations frequency ω_f. Note also that for nonpolar diatomic molecules both methods are equivalent, since, for this case, the external field does not perturb the potential, and we obtain the following equality: $\alpha_{V,J}^{UP} = \alpha_{V,J}^{PP}$.

3.1.1 Rovibrational Levels

The energy values of vibrational-rotational states $E_{V,J}^F$ for several diatomic molecules of general interest, namely, H_2, N_2, O_2, CO, NO, OH, CH, HF, and HCl, were found in our work [8] by solving Eq. (3.1) for the scaled potential energy curves $U_{scal}^F(r)$. As an example, Fig. 3.1 shows the values of $E_{V,J=0}^0$ for H_2, O_2, and CO obtained in [8] and reported elsewhere [11–14] as well as the fitting of spectroscopic data for H_2 [15] and O_2 [16]. Clearly, the methodology applied in [8] made it possible to adequately predict the energy of bound vibrational levels in diatomic molecules under consideration, which is not surprising given the good description of potential energy curves demonstrated in Fig. 2.1.

As for the rotational level structure, as can be seen from Fig. 3.2, it is also predicted for diatomics with reasonable accuracy, at least for low energies. Of course, if one wants to reproduce the position of all rovibrational levels of diatomic molecules more accurately, it is necessary, as noted above (see Sect. 2.1), to use higher-level quantum chemical methods than was done in [8]. The features of such calculations can be found elsewhere [17–19]. However, it is worth mentioning that obtaining spectroscopic accuracy for rovibrational energy levels $E_{V,J}^F$ is a task of particular intricacy and lies away from the core problems described in this book.

Fig. 3.1 Energy of vibrational levels $E_{V,J=0}^0$ of H_2, O_2, and CO against vibrational quantum number V calculated in [8] (curves) and elsewhere [11–16] (symbols) (Reused with permission from Ref. [8]. Copyright 2016 IOP Publishing Ltd)

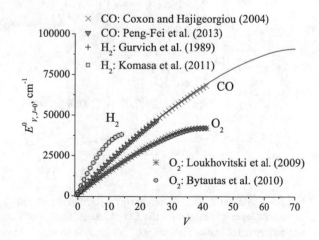

Fig. 3.2 Energy of
rovibrational levels $E^0_{V,J}$ of
H_2 as a function of rotational
quantum number J obtained
in [8] (curves) as well as the
reference data [15]
(symbols)

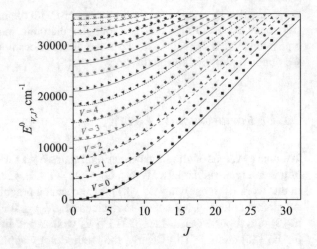

3.1.2 Dipole Moment

The state-specific values of effective dipole moment $\mu^F_{V,J}$ were calculated in [8] in compliance with Eq. (3.2) for all vibrational (V) and rotational (J) levels determined via solving Eq. (3.1). As an illustration of methodology, Fig. 3.3 portrays the potential $U^0(r)$ (superscript '0' stands for the absence of external field), energy of vibrational levels $E_{V,J=0}$ and corresponding radial wave functions $\chi^F_{V,J}(r)$ at $F=0$ for the CO

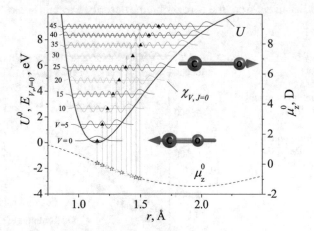

Fig. 3.3 Dependences of potential energy U^0 (solid curve) and dipole moment μ^0_z (dotted curve) on internuclear separation r for the CO molecule. Energy of vibrational levels $E^0_{V,J=0}$ and corresponding radial wave functions $\chi^0_{V,J}(r)$ (at $F=0$) together with the values of $\mu^0_{V,J=0}$ (stars, vertical arrows denote the corresponding vibrational level) and $r^0_{V,J=0}$ (closed triangles) for different vibrational states are also provided (Adapted with permission from Ref. [8]. Copyright 2016 IOP Publishing Ltd)

molecule. Also given there are the corresponding DMF together with the average values of interatomic distances $r^0_{V,J=0}$ for different vibrational quantum numbers (the values of $r^0_{V,J=0}$ are depicted at the respective energy levels). Stars, placed on the DMF curve, represent the effective dipole moment values $\mu^0_{V,J=0}$ of CO molecule for different vibrational states. As seen, the values of the effective dipole moment $\mu^0_{V,J=0}$ for higher vibrational levels correspond to $\mu^0_z(r)$ magnitudes at larger r. It is also essential that the value of $r^0_{V,J=0}$ is somewhat higher than the internuclear distance, at which the molecular dipole moment equals its averaged value (denoted by the arrows), notably, for high vibrational states. This observed difference stems from the fact that the DMF(r) dependence is almost linear near the r_e, but with greater stretching of the CO molecule, it falls less steeply.

Shown in Fig. 3.4 are the values of $\mu^0_{V,J=0}$ and $\mu^0_{V=0,J}$ as a function of V and J quantum numbers for different diatomic molecules. We clearly see that there is a fairly strong dependence of the dipole moment on the V number. However, the behavior of this dependence substantially differs for the diatomic molecules in question. For instance, for hydrogen halide molecules HCl and HF, the $\mu^0_{V,J=0}(V)$ dependence has a pronounced maximum at $V = 6$ and $V = 9$, respectively. Such behavior is owing to the fact that the maxima of $\mu^0(r)$ dependence for these molecules are located at notably longer internuclear distance than r_e (see Fig. 2.5), and vibrational stretching of these molecules leads to the increase in the effective dipole moment. With that, the dependence of dipole moment on the vibrational quantum number V for the OH radical has a faint maximum already at $V = 1$, and after that the observable dipole moment of OH decreases monotonically. Such behavior stems from the fact that maximum of $\mu^0(r)$ dependence for OH is attained at the r close to r_e (see Fig. 2.3a), and vibrational stretching of the molecule brings about a decrease in effective dipole moment.

In turn, state-specific dipole moments of CO and NO molecules diminish with V number, changing their signs at $V = 3$ and $V = 13$, respectively. Then, a further increase in V leads to the minimum (negative) values of $\mu^0_{V,J=0}$ at some $V = V_{min}$ (for CO, $V_{min} = 54$ and for NO, $V_{min} = 34$). Afterward, the $\mu^0_{V,J=0}$ values go back to zero at the largest V. Such behavior for CO and NO is caused by the sign change of their DMFs at a certain internuclear separation (see Figs. 2.3b and 2.4a). However, even though the DMF for the CH radical also reverses its sign (see Fig. 2.4b), the respective dependence of dipole moment on V is monotonic ($\mu^0_{V,J=0}$ steadily decreases with V increasing). This is because the maximal absolute value of the DMF for CH in the range where $\mu^0_z < 0$ is very small (-0.21 D).

The dependence of dipole moment of a diatomic molecule on its rotational quantum number is less dramatic (see Fig. 3.4b), but it is still noticeable for H-containing dipole molecules with large values of rotational quanta (HF, CH, HCl, and OH). The largest variation in dipole moment with J increasing is revealed for the CH molecule: the $\mu^0_{V=0,J=45}$ value for CH is 1.9 times smaller than the ground-state dipole moment $\mu^0_{V=0,J=0}$. Additionally, the calculated bivariate $\mu^0_{V,J}$ dependences for CH and HF molecules are given in Fig. 3.5. From the graphs we clearly see that the state-specific

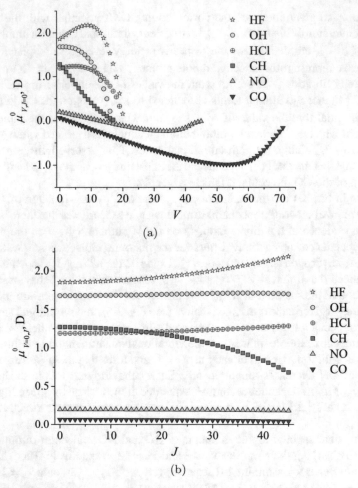

Fig. 3.4 Calculated values of $\mu_{V,J=0}^0$ (**a**) and $\mu_{V=0,J}^0$ (**b**) against the V and J numbers, respectively (Reused with permission from Ref. [8]. Copyright 2016 IOP Publishing Ltd)

effective dipole moment $\mu_{V,J}^0$ for CH is monotonic both in V and in J, whereas for HF the $\mu_{V,J}^0$ behavior, as expected (cf. Fig. 3.4), is more complex.

We next compare the obtained dependences with other past calculations. Presented in Fig. 3.6 are the $\mu_{V,J=0}^0 - \mu_{V=0,J=0}^0$ data predicted in [8] for CO, NO, and HF molecules and the findings of other researchers [20–27]. One can conclude that predictions of the effective dipole moment values for the given vibrational states [8] are in consistency with the theoretical calculations reported elsewhere.

Fig. 3.5 The calculated values of $\mu_{V,J}^0$ for CH (**a**) and HF (**b**) molecules

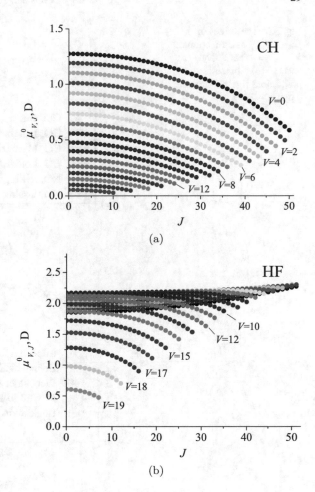

3.1.3 Dipole Polarizability

The state-specific values of polarizability $\alpha_{V,J}^{\mathrm{UP}}$ and $\alpha_{V,J}^{\mathrm{PP}}$ were calculated in [8] in line with Eqs. (3.3)–(3.4) and Eqs. (3.5)–(3.6) respectively for all vibrational-rotational states of molecules under consideration. As mentioned above (see Sect. 3.1), for non-polar diatomic molecules, the equality $\alpha_{V,J}^{\mathrm{UP}} = \alpha_{V,J}^{\mathrm{PP}}$ is valid. The comparison of $\alpha_{V,J}^{\mathrm{UP}}$ values obtained in [8] with the pioneering calculations of Kolos and Wolniewicz [28] and Rychlewski [29] for the H_2 molecule is shown in Fig. 3.7. One can discern that the predictions [8] for the state-specific polarizabilities are in sound agreement with the calculations [28, 29] for different vibrational and rotational levels. It was also found that the polarizability values for the H_2 molecule in its highest vibrational ($V = 12$, $J = 0$) and rotational ($J = 32$, $V = 0$) states are close together. Additionally, the calculated bivariate $\alpha_{V,J}^{\mathrm{UP}}$ dependence for H_2 is given in Fig. 3.8. From this

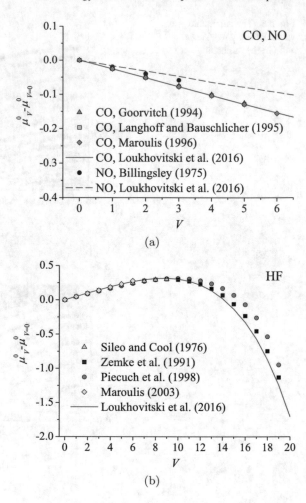

Fig. 3.6 The values of $\mu_V^0 - \mu_{V=0}^0$ as a function of V at $J = 0$ for CO, NO (**a**), and HF (**b**) molecules: predictions [8] and the other data [20–27]

graph we see that the state-specific polarizability for H_2 is, in whole (for not too large V numbers), monotonic both in V and in J.

It is worth pointing out that the alteration in the H_2 polarizability when it is excited from $V = 0$ to $V = 1$ was measured accurately using the optical Stark spectroscopy by Dyer and Bischel [30] almost three decades ago. For a long-wave limit, Dyer and Bischel [30] concluded that the value of $\Delta\alpha_{10} = \alpha_{V=1, J=0}^{\mathrm{UP}} - \alpha_{V=0, J=0}^{\mathrm{UP}}$ was equal to 0.067 Å^3. Our calculations [8], in turn, give $\Delta\alpha_{10} = 0.066$ Å^3 that corresponds well with the measurements and indicates the success of the applied approach.

As mentioned above, for polar diatomic molecules the external electric field perturbs the potential, and hence $\alpha_{V,J}^{\mathrm{UP}} \neq \alpha_{V,J}^{\mathrm{PP}}$. Shown in Fig. 3.9 are the values of $\alpha_{V, J=0}^{\mathrm{UP}}$ and $\alpha_{V, J=0}^{\mathrm{PP}}$ for the HF molecule, predicted in [8], and those calculated for $V = 0..5$ by Marti et al. [31] and by Maroulis [27]. One may see that the $\alpha_{V, J=0}^{\mathrm{PP}}$ values are only negligibly higher than the $\alpha_{V, J=0}^{\mathrm{UP}}$ ones for $V < 12$, whereas, for higher vibra-

Fig. 3.7 The values of $\alpha^{UP}_{V,J=0}$ and $\alpha^{UP}_{V=0,J}$ as a function of V and J numbers for the H_2 molecule as predicted in [8] (solid and dashed curves, respectively) and calculated by Kolos and Wolniewicz [28] and Rychlewski [29] (symbols) (Reused with permission from Ref. [8]. Copyright 2016 IOP Publishing Ltd)

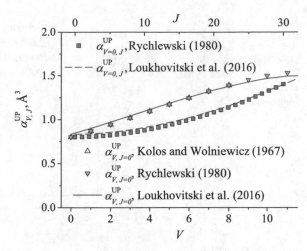

Fig. 3.8 The calculated values of $\alpha^{UP}_{V,J}$ for H_2 molecule (Reused with permission from Ref. [8]. Copyright 2016 IOP Publishing Ltd)

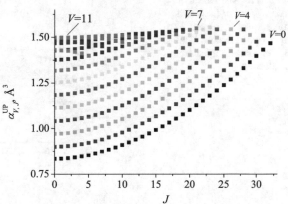

tional states, the $\alpha^{PP}_{V,J} / \alpha^{UP}_{V,J}$ ratio can be as large as 1.5. It should be recalled (see the beginning of this Section) that the PP method is relevant to the static electric field, whereas for the alternating electric field, the effect of perturbation of vibrational wave function by the external electric field disappears with increasing the frequency of electric field oscillations ω_f. Thus, as observed from Fig. 3.9, the neglect of the effect of a probe field on the vibrational wave function can result in a remarkable underestimation of the polarizability of polar diatomic molecules at the static case, especially for the upper vibrational levels.

For definiteness, we illustrate further the effect of multiple vibrational excitation on polarizability precisely for the PP case. The values of $\alpha^{PP}_{V,J=0}$ and $\alpha^{PP}_{V=0,J}$ with different V and J numbers for diatomic molecules at issue calculated in [8] are given in Fig. 3.10. From this, we may ascertain that the magnitude of state-specific polarizability strongly depends on the vibrational quantum number. This implies that, by exciting the vibrations of a molecule, we can notably change its polarizability. The $\alpha^{PP}_{V,J=0}$ values for diatomic molecules excited to high vibrational states can be

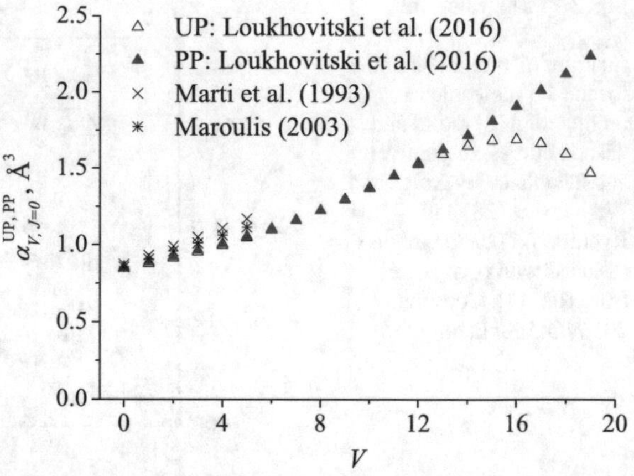

Fig. 3.9 The values of $\alpha_{V,J=0}^{UP}$ and $\alpha_{V,J=0}^{PP}$ as a function of V number for the HF molecule calculated in [8] and predicted by Marti et al. [31] and by Maroulis [27]

Fig. 3.10 The values of $\alpha_{V,J=0}^{PP}$ and $\alpha_{V=0,J}^{PP}$ as a function of V (**a**) and J (**b**) quantum numbers calculated in [8] for certain diatomic molecules

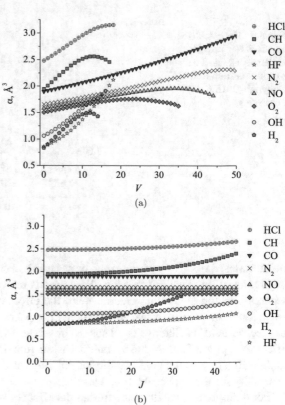

1.5–3.0 times greater than those for nonexcited ones (see Fig. 3.10a). With that, the influence of rotational excitation on the observable polarizability is less marked (see Fig. 3.10b). Even for the lightest H_2 and CH molecules, the relative increase in their polarizabilities does not exceed two times when exciting molecules to high rotational states ($J > 30$).

It is worth mentioning that the contribution of the nuclear zero-point motion to the molecular polarizability, that is commonly specified in the related papers as the zero-point vibrational averaged polarizability $\Delta\alpha^{ZPV}$ [10, 32], can be cast in terms of the analysis [8] as

$$\Delta\alpha^{ZPV} = \alpha^{UP}_{V=0, J=0} - \alpha^{UP}(r_e). \tag{3.7}$$

For polar molecules, some researchers also introduce the term "pure vibrational polarizability" $\Delta\alpha^{PV}$. It is customary to associate this term with the fact that the external static field disturbs vibrational wave function, which makes an additional contribution to the observed polarizability in comparison with the effect of the nuclear zero-point motion [10, 33, 34]. In the framework of the methodology under consideration, upon the $\alpha^{PP}_{V,J}$ estimating, this effect is taken into consideration implicitly, as the vibrational wave functions, in the case in question, are calculated for the internuclear potential perturbed by the applied electric field. Via the quantities previously introduced, the magnitude of $\Delta\alpha^{PV}$ is stated as follows

$$\Delta\alpha^{PV} = \alpha^{PP}_{V=0, J=0} - \alpha^{UP}_{V=0, J=0}. \tag{3.8}$$

In this regard, it is worth comparing the values of $\Delta\alpha^{ZPV}$ and $\Delta\alpha^{PV}$ predicted based on Eqs. (3.7) and (3.8) with those reported by other researchers [28, 32, 35–40]. The respective data are listed in Table 3.1. It can be seen that these quantities are fairly

Table 3.1 The values of $\Delta\alpha^{ZPV}$ and $\Delta\alpha^{PV}$ corrections for H_2, N_2, O_2, CO, HF, and HCl molecules reported elsewhere and estimated in [8] (units are $Å^3$)

Molecule	$\Delta\alpha^{ZPV}$		$\Delta\alpha^{PV}$	
	[8]	Other works	[8]	Other works
H_2	0.037	0.035 [28]	–	–
N_2	0.021	0.007 [35]	–	–
		0.007 [36]	–	–
O_2	0.016	0.005 [37]	–	–
		0.008 [38]	–	–
CO	0.006	0.007 [35]	0.020	0.018 [35]
HF	0.013	0.013 [32]	0.010	0.010 [32]
		0.015 [35]		0.009 [35]
		0.015 [39]		
HCl	0.034	0.021 [32]	0.004	0.006 [32]
		0.022 [40]		0.007 [40]

small for considered diatomic molecules [$\Delta\alpha^{\text{ZPV}} < 0.05\,\alpha(r_e)$, $\Delta\alpha^{\text{PV}} < 0.02\,\alpha(r_e)$], although of course, the inclusion of vibrational effects to dipole polarizabilities of diatomic molecules is undoubtedly desirable when comparing the results of very accurate theoretical calculations and precision measurements. From a comparison with the data [28, 32, 35–40] listed in Table 3.1 we can also infer that the methodology made use of in [8] and considered here allows us, on the whole, to predict correctly the effect of zero-point oscillations on the polarizability of diatomic molecules.

3.2 Polyatomic Molecules

The main trends of the influence of vibrational nuclear motion in polyatomic molecules and clusters on their electric properties were considered in our work [41] for a set of 50 compounds, significant for combustion, plasma, atmospheric, and material chemistry. This representative set covers different molecules (small hydrocarbons, H_xO_y and $H_xC_yO_z$ species together with several nitrogen-, silicon-, sulfur-, and halogen-containing molecules) and small atomic covalent clusters (involving C, Al, B, Si, P, S, some alkaline and alkaline-earth elements). Only the ground electronic states were assigned for all compounds, and atomic clusters were taken in their lowest energy isomeric forms. Note also that prior to the calculations of electric properties, all structures were optimized at the UB97-2/Sadlej or (if necessary) UB97-2/aug-cc-pVTZ levels of treatment.

To evaluate the dipole moment and static polarizability averaged over a vibrational motion of particular mode m, the corresponding vibrational wave functions $\chi_m^F(r)$ were obtained within the Born-Oppenheimer approximation via solving the steady-state Schrödinger equation for each normal mode. The latter is expressed in the conventional form

$$\frac{d^2\chi_m^F(r)}{dr^2} + \left[\frac{2M_m}{\hbar^2}\left(E - U_m^F(r)\right)\right]\chi_m^F(r) = 0, \tag{3.9}$$

where M_m is the reduced mass for a given vibrational mode, E is the energy of vibrational motion (cf. Eq. (3.1)). We note here that the rotational motion of the molecule was ignored in [41] and, accordingly, hereinafter. Equation (3.9) was numerically solved for four cases mentioned previously ($F = 0$, X, Y, Z, see also Sect. 2.2) using the approach by Numerov [2, 3]. Eventually, eigen energy values $E_{m,V}^F$ together with corresponding wave functions $\chi_{m,V}^F(r)$ for a given vibrational quantum number V of the mth mode were determined.

Absolute magnitudes of effective dipole moment $\mu_{V_1...V_N}$ for a specific vibrational state $(V_1...V_N)$ were determined through the respective vector components as follows:

$$\mu_{V_1\ldots V_N,i} = \mu_{e,i} + \sum_{m=1}^{N} \Delta\mu_{m,i}, \, i = x, \, y, \, z,$$

$$\Delta\mu_{m,i} = \langle \chi^0_{m,V_m} | \mu^0_{m,i}(r) | \chi^0_{m,V_m} \rangle - \mu_{e,i}.$$

(3.10)

Here $\mu_{e,i}$ are the components of the vectorial dipole moment at the PES minimum, and $\Delta\mu_{m,i}$ are the contributions of each vibrational mode to effective dipole moment.

Similar to diatomic molecules in [8], the level values of polarizability were calculated through the use of two different approaches rooted in the finite-field method [5, 6]: UP and PP (see Sect. 3.1). In the UP case (the impact of the probe field on the potential is insignificant), the dependence of polarizability on the normal vibrational coordinate for the mth mode $\alpha_m^{UP}(r)$ is stated as follows:

$$\alpha_m^{UP}(r) = \left(\alpha_m^{UP,1}(r) + \alpha_m^{UP,2}(r) + \alpha_m^{UP,3}(r)\right)/3,$$

(3.11)

where $\alpha_m^{UP,1}(r)$, $\alpha_m^{UP,2}(r)$, and $\alpha_m^{UP,3}(r)$ are the diagonal elements of polarizability tensor, built, in turn, by diagonalizing the tensor with the following terms:

$$\alpha_{m,i,j}^{UP}(r) = \frac{\mu_{m,i}^j(r) - \mu_{m,i}^0(r)}{\mathcal{E}}, i = x, \, y, \, z; \, j = X, \, Y, \, Z.$$

(3.12)

Here $\mu_{m,i}^j(r)$ is the ith component of dipole moment vector, when the external electric field of strength $\mathcal{E}$ is applied along the axis Oj, $\mu_{m,i}^0(r)$ is the ith component of dipole moment vector in the absence of external electric field. Consequently, the state-specific polarizability values $\alpha_{V_1\ldots V_N}^{UP}$ are stated as

$$\alpha_{V_1\ldots V_N}^{UP} = \alpha_e + \sum_{m=1}^{N} \Delta\alpha_{m,V_m}^{UP},$$

$$\Delta\alpha_{m,V_m}^{UP} = \langle \chi^0_{m,V_m} | \alpha_m^{UP}(r) | \chi^0_{m,V_m} \rangle - \alpha_{el},$$

(3.13)

where α_{el} is polarizability at the PES minimum (known as electronic polarizability), and $\Delta\alpha_{m,V_m}^{UP}$ are the contributions of different modes to an effective polarizability.

The second approach to calculate the polarizability allows for the perturbation of potential by an external probe field (PP). Following this, the values of state-resolved polarizability are determined as

$$\alpha_{V_1\ldots V_N}^{PP} = \alpha_{el} + \sum_{m=1}^{N} \Delta\alpha_{m,V_m}^{PP},$$

$$\Delta\alpha_{m,V_m}^{PP} = \left(\alpha_{m,V_m}^{PP,1}(r) + \alpha_{m,V_m}^{PP,2}(r) + \alpha_{m,V_m}^{PP,3}(r)\right)/3 - \alpha_{el},$$

(3.14)

where $\Delta\alpha^{PP}_{m,V_m}$ are the contributions from different modes to an effective polarizability, $\alpha^{PP,1}_{m,V_m}(r)$, $\alpha^{PP,2}_{m,V_m}(r)$, and $\alpha^{PP,3}_{m,V_m}(r)$ are the diagonal elements of polarizability tensor, determined by diagonalizing tensor with following elements:

$$\alpha^{PP}_{m,V_{m,i,j}}(r) = \frac{\langle \chi^j_{m,V_m} | \mu^j_{m,i}(r) | \chi^j_{m,V_m} \rangle - \langle \chi^0_{m,V_m} | \mu^0_{m,i}(r) | \chi^0_{m,V_m} \rangle}{\mathcal{E}}, \qquad (3.15)$$
$$i = x,\, y,\, z;\; j = X,\, Y,\, Z.$$

Notice that the contributions of the nuclear zero-point vibrational (ZPV) motion ($V_1 = 0, ..., V_N = 0$) to the dipole moment and polarizability of a polyatomic molecule (cluster) are routinely called in related literature as the zero-point vibrational averaged corrections to dipole moment $\Delta\mu^{ZPV}$ [42] and polarizability $\Delta\alpha^{ZPV}$ [10, 34]. They can be expressed in terms of [41] (cf. Eq. 3.7 for diatomics) as

$$\Delta\mu^{ZPV} = \mu_{V_1=0...V_N=0} - \mu_e, \qquad (3.16)$$

$$\Delta\alpha^{ZPV} = \alpha^{UP}_{V_1=0...V_N=0} - \alpha_{el}. \qquad (3.17)$$

The so-called "pure vibrational polarizability" $\Delta\alpha^{PV}$ usually introduced for polyatomic molecules by analogy with diatomics [43–46] within the employed framework is treated implicitly when estimating the $\alpha^{PP}_{V_1...V_N}$, inasmuch as the vibrational wave functions, in this case, are calculated for the potential modified by the applied electric field. Following from this, in terms of [41], the magnitude of $\Delta\alpha^{PV}$ is given by (cf. Eq. (3.8) for diatomics)

$$\Delta\alpha^{PV} = \alpha^{PP}_{V_1=0...V_N=0} - \alpha^{UP}_{V_1=0...V_N=0}. \qquad (3.18)$$

The PP method is relevant to the static field limit (the frequency of field oscillations ω_f is close to zero) when both types of vibrational corrections are to be reckoned with:

$$\alpha(\omega_f = 0) = \alpha_{el} + \Delta\alpha^{ZPV} + \Delta\alpha^{PV}. \qquad (3.19)$$

This occurs for the dispersion interaction of molecules and molecules placed in a radiofrequency field. However, as ω_f increases, the pure vibrational contribution $\Delta\alpha^{PV}$ tends to zero (so long as the effect of disturbance of vibrational wave function by the probe field becomes negligible), and this term may be disregarded under UV and visible radiation [34, 35]:

$$\alpha(\omega_f > 0.05 \text{ a.u.}) = \alpha_{el} + \Delta\alpha^{ZPV}. \qquad (3.20)$$

Thereby, importantly, the $\Delta\alpha^{PV}$ magnitude can be evaluated via a difference between the polarizabilities obtained from the static or radio-frequency measurements of dielectric permittivity (DP) and from optical measurements of the refractive index (RI). Notice that alternatively, the pure vibrational polarizability can be determined directly from the observations of the radiation intensity of the infrared-active bands [43, 44, 47].

To find the pure electronic part of dipole moment μ_e and polarizability α_{el} more accurately than can be implemented with the popular B97-2 functional, the refined post-Hartree-Fock calculations were employed in [41] in terms of the finite-field formalism. To this end, for the structures optimized at the B97-2/Sadlej DFT model, the MP4(SDQ) variant of the fourth-order Møller–Plesset perturbation theory [48] capable to reproduce the reliable polarizability values for a broad range of polyatomic species [49–51] was applied. The common notation for this computational model is MP4(SDQ)/Sadlej // B97-2/Sadlej.

3.2.1 Vibrational Levels

Now we turn to a discussion of the basic results obtained in [41] within the methodological framework outlined above. To begin with, of distinct interest is the validity of the approach applied in [41] for exploring the PES of polyatomic molecules. Let us confront the energy of vibrational levels, obtained utilizing the employed methodology, with the spectroscopic and accurate theoretical data. Displayed in Fig. 3.11 is the comparison of the energy of vibrational levels obtained in [41] with the literature data [15, 52] for the bending (ν_1), symmetric stretching (ν_2), and asymmetric stretching (ν_3) modes of H_2O molecule, as well as for the bending (ν_{1-2}), symmetric stretching (ν_3), and asymmetric stretching (ν_4) modes of CO_2 molecule (combined vibrational states are not presented here). Note, in passing, that in [41] and within the present book, the molecular modes are numbered in order of increasing normal frequency, and degenerated modes are handled as distinct ones.

It can be seen that the energies of the low vibrational levels coincides with the reference data. This implies that the typical PESs, at least, near their minima, are reproduced with the help of the B97-2/Sadlej level of theory properly. However, the applied methodology starts to overstate the energetic levels with $E_V - E_{V=0} > 5000 \text{ cm}^{-1}$. This is due to the approximate nature of the assumptions made: (i) the rectilinear movement of atoms in a molecule along the lines of the displacement vectors of normal vibrations and (ii) treatment of vibrations through a system of separable one-dimensional oscillators instead of considering the motion in a single multidimensional potential well. Generally speaking, the use of curvilinear coordinates to describe the motion of atoms (nuclei) in a molecule can improve the consistency with accurate E_V values (see [53, 54]), but for very high levels, the vibrational mode coupling that was neglected in [41] becomes essential, and the approach grounded on independent oscillators cannot, in principle, yield reasonable E_V values. Indeed, the high (spectroscopic) accuracy in predicting the energy of vibrational states can be achieved only during rigorous full-dimensional ab initio calculations at quite sophisticated levels of treatment taking into account relativistic, spin-orbit coupling, and, perhaps, nonadiabatic corrections [17, 55].

Fig. 3.11 The position of vibrational levels of different modes of H₂O (**a**) and CO₂ (**b**) molecules as a function of a vibrational quantum number V obtained in [41] (lines) as well as the respective data reported elsewhere for H₂O [52] and CO₂ [15] (symbols) (Reused with permission from Ref. [41]. Copyright 2017 IOP Publishing Ltd)

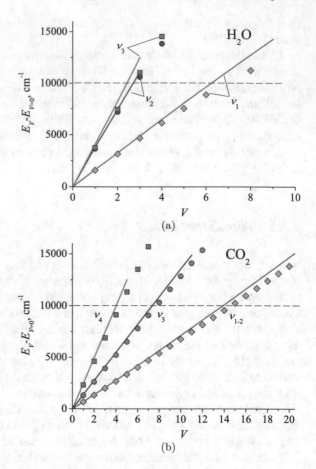

It also follows from Fig. 3.11 that the low-lying levels of H₂O and CO₂ triatomic molecules with the vibrational energy within the cutoff criterion $E_{cut} = 10000$ cm^{-1} can be reproduced by using the suggested methodology [41] with the accuracy no worse than 10%. Elementary thermodynamic estimates indicate that this magnitude of E_{cut} provides a proper evaluation of the temperature-dependent electric properties up to the gas equilibrium temperature $T \approx 3000$ K (at this temperature the aggregate population of vibrational states with $E_{vib} > E_{cut}$ does not surpass 1%).

3.2.2 Dipole Moment

To exemplify the performance of methodology described above and put in practice in [41], let us take a close look at the representative dependences of dipole moment and potential energy increments against the coordinate of normal vibrations for the

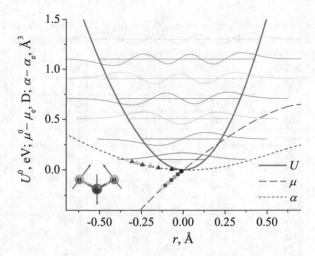

Fig. 3.12 The increments in dipole moment (dashed curve), polarizability (dotted curve), and potential energy (solid curve) versus the normal vibrational coordinate for the bending mode of H_2O molecule. The vibrational levels (horizontal lines) and corresponding vibrational wave functions (oscillating curves) are also represented for $V = 0..6$. The state-specific dipole moments and polarizabilities are provided by markers placed at the appropriate curves (Adapted with permission from Ref. [41]. Copyright 2017 IOP Publishing Ltd)

bending mode of H_2O molecule (see Fig. 3.12). For clarity, the energy levels, corresponding wave functions, and effective dipole moment values are presented there as well. As observed, the potential U^0 is somewhat asymmetric, as a result of which the vibrational coordinate upon averaging over the vibrational wave function tends, with the V number growth, to shift towards the branch of potential with a shallower gradient. So long as the $\mu^0(r)$ dependence is almost linear near the equilibrium point and has a positive slope, the displacement along effective coordinate brings about lowering the state-specific dipole moments with V increase.

The pure electronic part of dipole moment μ_e for polar polyatomics at issue was calculated applying both B97-2 and MP4(SDQ) theoretical models. The comparison of derived μ_e values for some molecules of general interest with the various measurements [56–59] is provided in Table 3.2. The contribution of ZPV motion to the molecular dipole moment $\Delta\mu^{ZPV}$ is given there as well. One can observe that both B97-2 and MP4(SDQ) μ_e predictions are fairly close to the experimental one, but the MP4(SDQ) method yields somewhat better consistency with the measurements (mean unsigned deviations from the reference μ_e values at these computational levels are 0.09 and 0.08 D, respectively). You may see that the use of correction for zero-point vibrational motion $\Delta\mu^{ZPV}$ enables one, by and large, to improve a little the agreement between the calculated and measured dipole moments (in this case, mean unsigned deviation from the experiment was reduced down to about 0.06 D).

Actually, even for large molecules (i.e. with many vibrational modes), the contribution of zero-point motion to the observable dipole moment is small as compared

Table 3.2 The values of μ_e, determined at the B97-2/Sadlej and MP4(SDQ)/Sadlej // B97-2/Sadlej computational levels, as well as the $\Delta\mu^{ZPV}$ magnitude for the molecules in question, estimated in [41], in comparison with the existing experimental data on the dipole moment (units are D)

Molecule	B97-2	MP4(SDQ)	$\Delta\mu^{ZPV}$	Total[a]	Experiment
H_2O	1.85	1.85	−0.01	1.84	1.85 [59]
HO_2	2.2	2.12	−0.02	2.1	2.09 [56]
O_3	0.65	0.62	0.0	0.62	0.53 [59]
H_2S	1.04	1.0	0.0	1.0	0.98 [59]
SO_2	1.75	1.8	0.0	1.80	1.63 [59]
HCN	3.01	3.03	0.0	3.03	2.99 [58]
NH_3	1.55	1.56	−0.07	1.48	1.47 [59]
NO_2	0.33	0.34	0.0	0.34	0.32 [59]
H_2O_2	1.78	1.78	−0.06	1.72	1.57 [59]
H_2CO	2.36	2.34	−0.05	2.29	2.33 [59]
C_3H_8	0.09	0.13	0.01	0.14	0.08 [59]
CH_3OH	1.61	1.65	−0.03	1.62	1.70 [59]
C_2H_5OH	1.54	1.71	−0.06	1.66	1.69 [59]
H_2SO_4	3.01	2.97	−0.01	2.96	2.73 [57]
Mean. Dev.	0.09	0.08	–	0.06	–

[a]MP4(SDQ)+$\Delta\mu^{ZPV}$

to the value of μ_e itself. Specifically, for the polar polyatomic molecules in question, the absolute magnitude of $\Delta\mu^{ZPV}$ correction is within 0.1 D. This is because the ZPV contributions from individual modes to the whole dipole moment are vectors, in effect, of various directions, partially counteracting each other so that the resulting ZPV vector is typically quite small in magnitude (see Eq. 3.10).

It should be pointed out that the accuracy in dipole moment computations for polyatomic molecules, achieved in [41] (∼0.06 D), is regarded as rather high, since the accuracy that can be reached for polyatomics using conventional quantum chemical methods and general-purpose basis sets is typically worse than 0.1 D [60, 61]. To improve this accuracy, the extra and often nontrivial efforts, such as a basis set extrapolation [62], special tailoring of basis sets [63–65], the property-oriented tuning of functionals [66–69], the use of CCSD(T) level of theory [70, 71] should be undertaken for the specific atomic system. Moreover, achieving the improved accuracy for dipole moments (∼10^{-3} D) is hardly feasible within the Born-Oppenheimer framework [72, 73].

We next examine the influence of exciting molecular vibrations on the dipole moment. In compliance with Eq. (3.10) the state-specific values of effective dipole moment of the molecules under consideration were calculated in [41] for the vibrational manifold within $E_{cut} = 10000$ cm^{-1}. As an example, Fig. 3.13 shows the μ_{v_100}, μ_{0v_20}, and μ_{00v_3} magnitudes for various modes of H_2O molecule (for bending, symmetric, and asymmetric stretching vibrations, respectively), calculated there and in [53, 74, 75] as well as registered in [76, 77]. We see here a reasonable correspon-

Fig. 3.13 The state-specific
values of dipole moments for
(V_100), $(0V_20)$, and $(00V_3)$
vibrational levels of bending,
symmetric, and asymmetric
stretching vibrations of H_2O
molecule, respectively:
predictions [41] (solid lines),
measurements [76, 77] and
calculations [53, 74, 75]
(symbols). The state-specific
values [53, 74, 75] were
shifted in such a way as to
match the experimental
magnitude [76] for (000)
level (Adapted with
permission from Ref. [41].
Copyright 2017 IOP
Publishing Ltd)

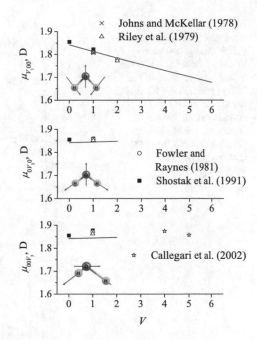

dence between calculations [41] and the data reported elsewhere. With that, one can observe that the excitation of bending mode of the H_2O molecule causes the decrease in effective dipole moment, while the excitation of symmetric and asymmetric stretch vibrations has a limited positive effect on the H_2O dipole moment.

A different sign of the effect of the excitation of specific modes on the dipole moment also holds for such molecules as H_2CO (formaldehyde) and HCN (hydrocyanic acid). As can be seen from Fig. 3.14, both the calculations of [41] and the measurements [78, 79] evidence that the excitation of CO-stretching vibrations gives rise to the growth of the H_2CO dipole moment. The excitation of out-of-plane bending vibrations, contrarily, results in its decrease (see Fig. 3.14a). A similar situation, as evidenced from Fig. 3.14b, holds also for HCN molecule. Interestingly, the superior coincidence between the predicted and measured values of state-specific dipole moment occurs for the stretching modes of H_2CO and HCN compounds rather than for the bending ones. This is so because for the former type of vibrations the assumption of the rectilinear movement of atoms along the displacement vectors of normal vibrations, employed in [41], is more justified.

From the plots displayed in Figs. 3.13 and 3.14 we see that the impact of vibrations on the dipole moment of polyatomic molecules is virtually rather small. Indeed, for vibrational levels below E_{cut}, the contribution of excitation of individual modes to the effective dipole moment of molecules at issue lies within the range 0.01–0.2 D (relative deviation compared to μ_e value does not go beyond 10%). This magnitude is comparable to the uncertainty inherent in the routine electronic structure calculations of molecular dipole moment ($\sim$0.1 D [60, 61]).

Fig. 3.14 Changes in the state-specific dipole moment values $\Delta\mu$ against the vibrational quantum number for each mode of H_2CO and HCN predicted in [41] (solid lines) and measured elsewhere [58, 78–80] (symbols) (Adapted with permission from Ref. [41]. Copyright 2017 IOP Publishing Ltd)

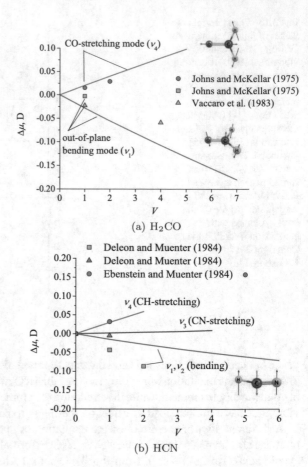

(a) H_2CO

(b) HCN

So far, the presentation was concerned with the polar molecules, for which asymmetry in the charge distribution takes place already in the equilibrium nuclear configuration. In this regard, it is interesting whether the vibrational motion can lead to the emergence of a dipole moment in the molecules that do not have a permanent dipole moment. For high-symmetry polyatomic molecules, usually thought of as nonpolar ones (such as CH_3, CH_4, C_2H_4, SiH_4, etc.), the magnitude of $\Delta\mu^{ZPV}$ correction obtained in [41] was found close to zero. In other words, this implies that for nonpolar molecules, at least, in their ground electronic states, ZPV motion cannot induce appreciable dipole moment. This conclusion correlates with the high-precision measurements and theoretical investigations [81–84], implying that such molecules can possess in their ground vibrational state only an exceedingly small dipole moment ($\sim 10^{-5}$ D). We note, in passing, that if the isotopic substitution breaks the molecular symmetry, the effect of ZPV motion on the observable dipole moment of such molecules can be more pronounced [85].

Fig. 3.15 Calculated values of state-specific dipole moments for the deformation modes of CH_3, CH_4, and Be_3 species versus V number (Adapted with permission from Ref. [41]. Copyright 2017 IOP Publishing Ltd)

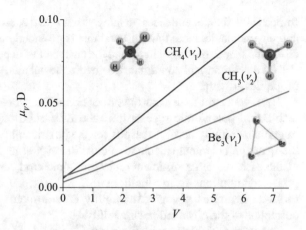

However, it should be stressed that the selective excitation of vibrations of certain modes in symmetric molecules can still lead to the emergence of nonzero dipole moment. This was predicted in [41] for the modes with either asymmetric $U_m^F(r)$ dependence or asymmetric $\mu_{m,i}^0(r)$ dependence (along, at least, one of the coordinate axes). Within the set of structures considered in [41], such modes are present, for instance, in the molecules of D_{3h} (CH_3, BH_3, AlH_3, Al_3, B_3, Be_3, Mg_3, Be_5) and T_d symmetry (Be_4, CH_4, SiH_4 etc.) Fig. 3.15 exemplifies this effect for the excitation of deformation vibrations of CH_3 (ν_2), CH_4 (ν_1), and Be_3 (ν_1) species. In this connection, it should be recalled that in the present book the molecular modes are enumerated in order of increasing frequency, and degenerated modes are handled as distinct ones. As observed, the excitation of specific modes in nonpolar compounds can induce a small but fundamentally nonzero dipole moment (up to ~ 0.1 D at $V > 5$). Although, it is worth recognizing that the selective excitation of high-lying vibrational states of individual modes in polyatomic molecules is very difficult to implement in practice by virtue of the strong coupling between the modes at high vibrational energy [86, 87].

3.2.3 Dipole Polarizability

To reveal the performance of methodology [41] as applied specifically to polarizability, let us look again at Fig. 3.12 (Sect. 3.2.2), where the representative dependences of polarizability on the coordinate of normal vibrations for the bending mode of H_2O molecule are depicted in addition to analogous dependences for potential energy increment and dipole moment. As can be seen, the situation for polarizability is slightly different from the situation for the dipole moment: the $\alpha^{UP}(r)$ dependence is almost parabolic with a minimum at the equilibrium geometry. Because the vibrational wave function, with V increase, tends to span over a broader range of vibrational coordinate, the certain growth of the state-specific polarizability occurs.

Importantly, the dependences, displayed in Fig. 3.12, are not typical for any modes and any molecule. Note that this example is provided only to demonstrate that precisely the shapes of the potential and the character of dependences of electric properties on the vibrational coordinate do govern the behavior of state-specific molecular properties at issue.

Next, we begin to discuss the effect of nuclear vibrational motion on the polarizability of polyatomic species from the ZPV contribution. It is well known that to attain extremely accurate results for the polarizability of polyatomic molecules, the quantum chemical calculations must involve, alongside a scrupulous selection of the basis set and a sufficient inclusion of electron correlation effects, the appropriate contributions from vibrations [34, 88]. Even if one would inquire into the room-temperature value of polarizability, the zero-point motion can cause a sizeable correction of the observed polarizability.

The values of ZPV corrections $\Delta\alpha^{ZPV}$ and $\Delta\alpha^{PV}$ were calculated in [41] for considered molecules based upon Eqs. (3.17) and (3.18), as described above. The found values of $\Delta\alpha^{ZPV}$ and $\Delta\alpha^{PV}$ are interesting to confront with these, reported by other researchers elsewhere based on calculations [31, 39, 40, 42, 45, 46, 89, 91–100] and the measured intensities of the molecular infrared-active bands [44, 47]. The respective data are listed in Table 3.3. We can infer that in general, the methodology applied in [41] makes it possible to predict the effect of zero-point motion on the polarizability of polyatomic molecules properly (aside from O_3). The substantial underestimate of $\Delta\alpha^{ZPV}$ magnitude with respect to the findings of Naves et al. [46] for O_3 molecule (by a factor of 15) is supposedly because of its severe multireference nature [101]. In fact, the DFT methodology applied in [41] is not capable to describe adequately the electronic structure of the ground state ozone [102], whereas the results of Naves et al. [46] were grounded on the CC method, more relevant for such systems with important nondynamical correlation effects.

Besides, the remarkable discrepancy between calculations [41] and the data by Santiago et al. [95] is found for the $\Delta\alpha^{PV}$ magnitude of hydrogen peroxide H_2O_2 molecule. Both works indicate that for H_2O_2 $\Delta\alpha^{PV} \gg \Delta\alpha^{ZPV}$, but in the study by Santiago et al. [95] substantially higher value of $\Delta\alpha^{PV}$ is provided. Conceivably, the value [95] (2.8 Å^3) is more accurate, since Santiago et al. [95] implemented a special handling of H_2O_2 torsion mode, whereas the methodology [41] implies only the rectilinear atomic displacements during the vibrational motion.

From Table 3.3 one can observe that the values of $\Delta\alpha^{ZPV}$, being rather small for triatomic molecules (<0.1 Å^3), can be considerably greater for larger species (0.1-0.4 Å^3). Meanwhile, for some compounds, the $\Delta\alpha^{PV}$ correction may exceed $\Delta\alpha^{ZPV}$ one by an order of magnitude. In general, we can say that these contributions are sizeable enough to explain numerous discrepancies between experimental polarizability data and top-level calculations of α_{el} [60].

In Fig. 3.16 you may see the calculated α_{el} [of course, at the MP4(SDQ) level of theory], $\alpha_{V_1=0...V_N=0}^{UP}$, and $\alpha_{V_1=0...V_N=0}^{PP}$ values compared to the polarizabilities determined based on the RI and DP at room temperature [47, 103–105]. As observed, the usage of $\Delta\alpha^{ZPV}$ permits one to enhance insignificantly the agreement with polarizability values, determined through the refractive index (in particular, for H_2O_2, CH_4,

Table 3.3 The values of $\Delta\alpha^{ZPV}$ and $\Delta\alpha^{PV}$ for the molecules under study, estimated in [41] and reported by other researchers (units are Å^3)

Molecule	$\Delta\alpha^{ZPV}$		$\Delta\alpha^{PV}$	
	[41]	Other works	[41]	Other works
Triatomic molecules				
H_2O	0.026	0.040 [89]	0.043	0.037 [44]
		0.042 [42]		0.04 [90]
		0.042 [45]		0.045 [45]
		0.046 [91]		
		0.043 [92]		
CO_2	0.017	0.004 [93]	0.312	0.26 [44]
				0.289 [94]
				0.323 [47]
O_3	0.013	0.216 [46]	0.208	0.209 [46]
H_2S	0.042	0.059 [40]	0.0	0.001 [40]
SO_2	0.013	–	0.338	0.297 [44]
HCN	0.033	0.036 [89]	0.140	0.144 [47]
				0.21 [90]
Na_3	0.135	–	2.543	1.72 [90]
Tetra-atomic molecules				
H_2O_2	0.061	0.039 [91]	0.592	2.8 [95]
		0.027 [95]		
NH_3	0.046	0.081 [42]	0.187	0.234 [44]
		0.078 [39]		
H_2CO	0.080	–	0.094	0.062 [44]
C_2H_2	0.088	–	0.395	0.449 [47]
				0.48 [90]
Penta-atomic molecules				
CH_4	0.099	0.123 [96]	0.035	0.034 [47]
		0.123 [31]		0.041 [97]
		0.130 [42]		0.04 [90]
		0.124 [39]		
		0.133 [97]		
SiH_4	0.123	0.180 [40]	0.674	0.775 [40]
				0.566 [47]
CCl_4	0.05	0.08 [97]	1.138	0.692 [47]
				0.973 [97]
CF_4	0.018	–	1.047	0.919 [47]
Molecules with more than 5 atoms				
C_2H_4	0.103	0.150 [31]	0.134	0.186 [98]
		0.154 [98]		0.135 [47]
		0.138 [99]		
SF_6	0.023	–	1.893	2.035 [47]
				2.29 [90]
C_2H_6	0.220	0.224 [100]	0.048	0.057 [100]
				0.049 [47]
C_3H_8	0.389	–	0.069	0.059 [47]

Fig. 3.16 The values of α_{el} (dotted lines), α^{UP} (dashed lines), and α^{PP} (solid lines) for the ground vibrational states of a number of molecules, calculated in [41], as well as the α values determined at room temperature via the refractive index (RI, open symbols) and dielectric permeability (DP, closed symbols) based on the data [47, 103–105] (Reused with permission from Ref. [41]. Copyright 2017 IOP Publishing Ltd)

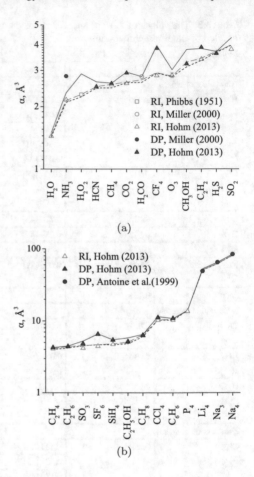

(a)

(b)

H_2CO, C_2H_6, and C_2H_5OH species). With that, it is the effect of the disturbance of molecular PES by external field (allowed for in $\alpha^{PP}_{V_1=0\ldots V_N=0}$ values) that makes it possible to properly interpret the difference between the DP and RI polarizability measurements (being particularly substantial for NH_3, CO_2, CF_4, C_2H_2, and SF_6 species).

We next discuss the effect of vibrational excitation on the polarizability of poly-atomic molecules. The values of the vibrational contributions of different modes $\Delta\alpha^{UP}_{m,V_m}$ and $\Delta\alpha^{PP}_{m,V_m}$ were calculated in [41] for all considered species in line with Eqs. (3.13) and (3.14) for all vibrational levels within $E_{cut} = 10000$ cm^{-1}. As mentioned previously, the state-specific polarizability calculations for polyatomic compounds are often limited by the effect of ZPV motion, therefore any data obtained for higher vibrational states (with separate contributions of individual vibrational modes) are of great significance.

Figure 3.17 displays the state-resolved values of $\Delta\alpha^{UP}_{m,V_m}$ and $\Delta\alpha^{PP}_{m,V_m}$ for the modes of H_2O molecule. The largest contribution to the polarizability according

Fig. 3.17 Contributions of normal modes to the vibrational polarizability corrections $\Delta\alpha_{m,V_m}^{UP}$ (**a**) and $\Delta\alpha_{m,V_m}^{PP}$ (**b**) for H_2O molecule (Adapted with permission from Ref. [41]. Copyright 2017 IOP Publishing Ltd)

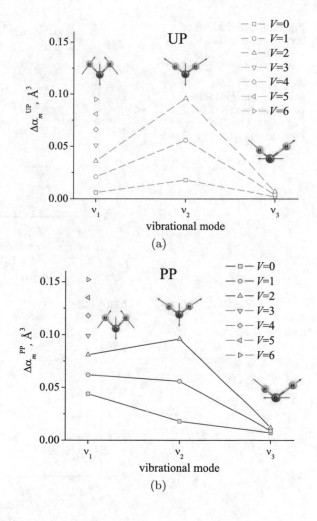

(a)

(b)

to the UP method comes from mode ν_2 (symmetric stretch), while the bending mode ν_1 also contributes substantially to the $\alpha_{V_1\ldots V_N}^{PP}$. Meanwhile, the effect of asymmetric stretching (ν_3) is small in both cases. We can ascertain that the $\Delta\alpha_{m,V_m}^{UP}$ and $\Delta\alpha_{m,V_m}^{PP}$ values for H_2O, in full accordance with the expectations [42, 96, 106], depend approximately linearly on V_m.

Similar tendencies are observed for other molecules. Specifically, Fig. 3.18 depicts the vibrational contributions of each mode of CH_4 molecule. We can observe that vibrational excitation of asymmetric stretching (ν_6) and bending (ν_{1-3}) modes predominately contribute to the polarizability of methane, while the effect of asymmetric stretching (ν_{7-9}) is the weakest.

Fig. 3.18 Contributions of normal modes to the vibrational polarizability corrections $\Delta\alpha_{m,V_m}^{UP}$ (**a**) and $\Delta\alpha_{m,V_m}^{PP}$ (**b**) for CH_4 molecule (Adapted with permission from Ref. [41]. Copyright 2017 IOP Publishing Ltd)

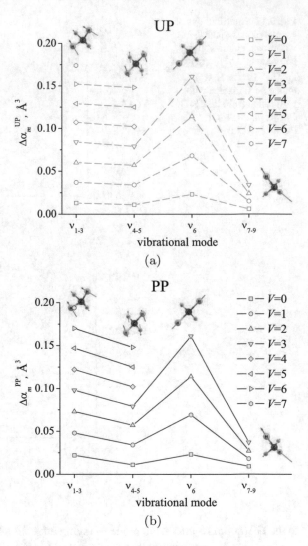

3.3 Approximation Formulae for the Effect of the Zero-Point Vibrations

Nowadays, theoretical methods of finding $\Delta\alpha^{ZPV}$ and $\Delta\alpha^{PV}$ corrections for small-sized molecules are well-known, readily available [8, 34, 41, 90] and even implemented into certain general-purpose quantum chemical packages [107]. However, their direct usage for the systems with plenty of electrons, atoms, and, consequently, vibrational degrees of freedom requires ample methodological and computational efforts. In particular, according to the approach [41], for each vibrational mode the proper $U_m^F(r)$ and $\mu_{m,i}^F(r)$ dependences are to be calculated. Virtually, such methods can hardly be customarily used for large molecules or clusters. Hence, the development of approximation schemes that enable us to estimate $\Delta\alpha^{ZPV}$ and $\Delta\alpha^{PV}$ magnitudes of isolated molecules or clusters via the quantities readily determined for the equilibrium structure of molecule (cluster) is of immense importance.

Recently, such a scheme based on the computed geometry, the values of α_{el}, μ_e, rotational constants, and vibrational frequencies was proposed [41]. Following the theory of vibrational contribution to polarizability for diatomics, $\Delta\alpha^{ZPV}$ magnitude is to be proportional to the ratio of corresponding characteristic rotational θ_{rot} and vibrational θ_{vib} constants [108, 109]. Additionally, the $\Delta\alpha^{ZPV}$ magnitude must depend on the α_{el} and internuclear separation r_e [10, 108]. As an analog of internuclear separation for polyatomic structures, the average distance to the center of mass R_{CM}, specifying the spatial extent and compactness of a particle, can be utilized here [110]. The latter for the system of S atoms is expressed by

$$R_{CM} = \frac{1}{S} \sum_i^S |\mathbf{r}_i - \mathbf{r}_{CM}|, \qquad (3.21)$$

where $\mathbf{r}_i$ is the vectorial position of ith atom and $\mathbf{r}_{CM}$ is the vectorial position of the center of mass. Assuming additivity of the contributions from vibrational modes [10], implied by the structure of Eqs. (3.13) and (3.14), the following three-parameter expression for approximating the results obtained numerically was proposed:

$$\frac{\Delta\alpha^{ZPV}}{\alpha_{el}} = p_0 \left(\frac{\alpha_{el}}{a_0{}^3} \right)^{p_1} \left(\frac{R_{CM}}{a_0} \right)^{p_2} \theta_{rot}^{red} \sum_m \frac{1}{\theta_{vib,m}}, \qquad (3.22)$$

$$\frac{1}{\theta_{rot}^{red}} = \begin{cases} \frac{2}{\theta_{rot}}, & \text{for linear molecules} \\ \sum_{i=1}^{3} \frac{1}{\theta_{rot,i}}, & \text{for nonlinear molecules} \end{cases}, \qquad (3.23)$$

where a_0 is the Bohr atomic radius, $\theta_{rot,i}$ are the characteristic rotational temperatures of molecule, θ_{rot}^{red} is a reduced rotational temperature, $\theta_{vib,m}$ is characteristic vibrational temperature of the mth mode, p_0, p_1, and p_2 are fitting parameters.

As for approximation of $\Delta\alpha^{PV}$ values, since the pure vibrational contribution is inherently related to the presence of permanent dipole moment [34, 44], the fitting expression is to be augmented with extra factor including dipole moment μ_e. With this in mind, the following five-parameter formula for $\Delta\alpha^{PV}$ approximation was proposed:

$$\frac{\Delta\alpha^{PV}}{\alpha_{el}} = p_3 \left(\frac{\alpha_{el}}{a_0{}^3} \right)^{p_4} \left(\frac{R_{CM}}{a_0} \right)^{p_5} \left(1 + \left(\frac{\mu_e}{p_6} \right)^{p_7} \right) \theta_{rot}^{red} \sum_m \frac{1}{\theta_{vib,m}}. \qquad (3.24)$$

The values of adjustable parameters $p_0..p_7$ were chosen to fit the results of $\Delta\alpha^{ZPV}$ and $\Delta\alpha^{PV}$ calculations for a representative set of 67 di- and polyatomic species

Table 3.4 The values of α_{el}, $\Delta\alpha^{ZPV}$, and $\Delta\alpha^{PV}$ for polyatomic molecules, calculated in [41] (units are Å^3), and the fitting parameters (b and c) for $\Delta\alpha^{UP}(T)$ and $\Delta\alpha^{PP}(T)$ (units are $\text{Å}^3\text{K}^{-1}$ and $\text{Å}^3\text{K}^{-2}$, respectively)

Species	α_{el}(B97-2)	α_{el}(MP4)	$\Delta\alpha^{ZPV}$	$\Delta\alpha^{PV}$	$b^{UP}\cdot 10^6$	$c^{UP}\cdot 10^9$	$b^{PP}\cdot 10^6$	$c^{PP}\cdot 10^9$
H_2O	1.444	1.430	0.026	0.043	-0.20	2.35	34.12	−6.69
HO_2	1.969	1.972	0.043	0.051	1.89	5.49	58.28	−9.72
CO_2	2.531	2.611	0.017	0.312	12.24	2.66	27.71	−4.07
O_3	2.727	2.801[a]	0.013	0.312	5.25	2.03	21.17	−2.88
H_2S	3.689	3.662	0.042	0.000	2.17	4.93	57.78	−10.27
SO_2	3.858	3.974	0.013	0.338	9.86	2.91	23.99	−2.22
NO_2	2.735	2.656	0.016	0.190	4.61	3.54	24.99	−2.42
SiO_2	4.433	4.577	0.027	1.956	52.52	1.76	−63.82	2.42
HCN	2.525	2.461	0.033	0.140	12.36	4.41	44.64	−6.67
H_2O_2	2.233	2.226	0.061	0.592	17.91	6.37	−46.57	4.34
CH_3	2.416	2.356	0.045	0.191	9.71	6.65	18.4	−3.39
NH_3	2.132	2.098	0.046	0.187	7.74	6.50	80.29	−11.46
H_2CO	2.645	2.621	0.080	0.094	6.24	9.98	111.39	−18.62
C_2H_2	3.475	3.352	0.088	0.395	45.66	6.68	126.18	−23.72
SO_3	4.383	4.506	0.018	0.724	25.57	4.46	39.54	−2.84
CH_4	2.507	2.464	0.099	0.035	10.05	12.72	140.07	−22.39
SiH_4	4.779	4.669	0.123	0.674	47.21	22.79	195.31	−24.4
CCl_4	10.435	10.278	0.050	1.138	82.25	11.10	154.22	−5.56
CF_4	2.851	2.907[a]	0.018	1.047	13.25	3.61	35.52	−3.05
C_2H_4	4.134	4.016	0.103	0.134	29.21	16.78	161.34	−20.95
CH_3OH	3.164	3.119	0.175	0.508	25.08	17.52	157.18	−31.54
H_2SO_4	5.553	5.047	0.130	2.738	88.40	10.41	−58.78	10.93
SF_6	4.682	4.665	0.023	1.893	38.18	5.43	57.94	−6.11
C_2H_6	4.294	4.207	0.220	0.048	71.23	23.22	360.02	−55.87
C_2H_5OH	4.997	4.827[a]	0.248	0.758	83.78	28.28	254.73	−36.35
C_3H_8	6.113	6.046[a]	0.389	0.069	132.70	34.43	642.69	−105
C_6H_6	10.231	10.292[a]	0.235	0.312	128.78	36.43	417.9	−48.43

[a] The triple substitutions were additionally taken into account during MP4 calculations

reported elsewhere. Specifically, the magnitudes of vibrational polarizability corrections for 17 diatomic molecules (OH [111]; Br$_2$ [112]; BH [7]; HBr [32]; HI [113]; LiH, NaH, KH [114]; BrCl [109]; H$_2$, N$_2$, O$_2$, CH, CO, NO, HCl, HF [8]) and 50 polyatomic species [41] were utilized for this purpose. In Tables 3.4 and 3.5 are listed the appropriate data as well as the calculated α_{el} values for molecules and clusters considered in [41]. Note that the $b^{UP/PP}$ and $c^{UP/PP}$ parameters in these tables will be described and discussed in the next Chapter.

Since the approximated quantities ($\Delta\alpha^{ZPV}$ and $\Delta\alpha^{PV}$) can vary by a few orders of magnitude, throughout the fitting, the root-mean-square deviation of decimal logarithms of target values is to be minimized. We note, in passing, that the nonpolar (homonuclear) diatomic molecules were eliminated from the $\Delta\alpha^{PV}$ approximation procedure, inasmuch as $\Delta\alpha^{PV} \equiv 0$ for them. The resulting values of fitting

Table 3.5 The values of α_{el}, $\Delta\alpha^{ZPV}$, and $\Delta\alpha^{PV}$ for clusters, calculated in [41] (units are Å^3), and the fitting parameters (b and c) for $\Delta\alpha^{UP}(T)$ and $\Delta\alpha^{PP}(T)$ (units are Å^{3}K^{-1} and Å^{3}K^{-2}, respectively)

Species	α_{el}(B97-2)	α_{el}(MP4)	$\Delta\alpha^{ZPV}$	$\Delta\alpha^{PV}$	$b^{UP} \cdot 10^6$	$c^{UP} \cdot 10^9$	$b^{PP} \cdot 10^6$	$c^{PP} \cdot 10^9$
Li$_3$	49.575	46.769	0.149	0.776	1289.59	−2.67	2514.47	−390.41
Be$_3$	19.168	19.258	0.063	0.026	163.94	−10.97	244.09	−33.25
Na$_3$	69.868	62.043	0.135	2.543	3754.43	−748.50	2502.07	−474.9
B$_3$[b]	7.693	9.986[c]	0.039	0.021	17.89	7.21	68.91	−6.61
Al$_3$[b]	24.241	25.236	0.149	0.118	613.43	−69.75	791.16	−120.49
Si$_3$	15.194	14.039	0.041	0.013	88.65	3.24	143.67	−11.67
S$_3$	10.192	6.076	0.021	0.371	48.00	−3.16	72.45	−11.29
Li$_4$	53.375	50.633	0.421	1.556	2839.39	−306.27	3179.7	−426.94
Be$_4$	20.471	21.149	0.084	0.228	116.24	20.77	223.41	−9.69
Na$_4$	79.608	80.720	0.203	2.822	3765.58	−578.44	4085.26	−698.38
Mg$_4$	44.311	44.123	0.085	0.186	611.25	−34.75	710.18	−63.72
B$_2$C$_2$[b]	6.686	6.605[a]	0.036	0.318	29.76	5.14	72.15	−7.38
B$_3$C[b]	7.874	8.219[a]	0.075	0.570	106.31	4.10	21.74	7.41
Al$_2$C$_2$[b]	16.785	16.644[a]	0.190	2.747	504.05	−42.62	795.59	−121.48
BH$_3$[b]	2.608	2.522	0.083	0.150	13.07	12.04	120.04	−18.41
AlH$_3$[b]	4.851	4.695	0.128	1.938	53.45	20.41	167.57	−30.89
AlB$_3$[b]	16.645	24.336[a]	0.080	0.762	119.23	15.17	214.71	−8.05
P$_4$	13.650	13.959[a]	0.011	0.005	39.22	5.54	53.87	1.49
Be$_5$	23.602	23.963	0.073	0.270	148.10	4.23	249.1	−24.46
C$_5$	10.498	11.311	0.120	2.367	230.18	−20.88	−735.61	193.48
B$_5$[b]	11.728	10.543	0.032	0.488	61.10	5.76	14.32	7.88
Al$_2$O$_3$[b]	8.838	8.727[a]	0.078	3.008	166.56	6.22	25.54	13.12
Al$_2$H$_4$[b]	11.759	11.538[a]	0.283	2.924	181.11	39.42	533.07	−67.84

[a] The triple substitutions were additionally taken into account during MP4 calculations
[b] Aug-cc-pVTZ basis set was used instead of the Sadlej pVTZ one
[c] The geometry was reoptimized at the MP4(SDQ)/aug-cc-pVTZ computational level

parameters were as follows: $p_0 = 36.875$, $p_1 = -0.734$, $p_2 = 1.332$, $p_3 = 85.069$, $p_4 = -1.504$, $p_5 = 5.312$, $p_6 = 4.721$ D, $p_7 = 2.161$.

Shown in Fig. 3.19 is the comparison of $\Delta\alpha^{ZPV}$ and $\Delta\alpha^{PV}$ values calculated rigorously in [41] and calculated by using Eqs. (3.22) and (3.24). Clearly, the application of Eq. (3.22) permits one to approximate most of the derived $\Delta\alpha^{ZPV}$ values with acceptable accuracy (the ratio between the approximated and the exact values is within 2–3). Much worse accuracy was observed for only a few molecules (see Fig. 3.19a). The situation is quite different for $\Delta\alpha^{PV}$ quantity. For this case, the approximating expression (3.24) provides a markedly larger difference between the fitted and calculated values. Virtually, the application of expression (3.24) enables one to achieve only an order-of-magnitude guess for the $\Delta\alpha^{PV}$ contribution to polar-

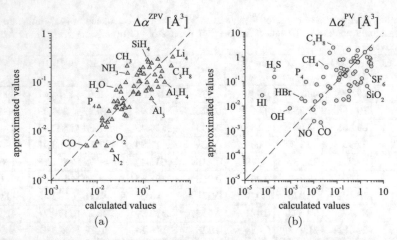

Fig. 3.19 The values of $\Delta\alpha^{ZPV}$ (**a**) and $\Delta\alpha^{PV}$ (**b**), approximated by Eqs. (3.22) and (3.24), versus the ones calculated in [41] (Reused with permission from Ref. [41]. Copyright 2017 IOP Publishing Ltd)

izability, while for certain molecules (they are HI, OH, and H_2S) this can lead to an overestimation by a factor of 10–100. Note that the fact that the worse fitting quality is found for $\Delta\alpha^{PV}$, apparently, stems from the peculiar nature of the pure vibrational contribution. As is recognized, the $\Delta\alpha^{PV}$ quantity is intrinsically related to a vibrational anharmonicity [91, 95], but the anharmonicity constants were not involved in the set of arguments in Eq. (3.24). It is possible that with the use of anharmonicity parameters a more accurate model for $\Delta\alpha^{PV}$ could be developed, however, the requirement to specify such properties (which are very difficult to estimate for large molecules) would inevitably narrow the field for the possible application of such approximation.

As seen from Tables 3.4 and 3.5, the set of polyatomic species at issue comprises the molecules and atomic clusters composed of, mainly, no more than 10 atoms. Hence, it would be pertinent in this context to verify the applicability of the elaborated approximations to systems of many atoms. For this purpose, we confronted the values, obtained using Eqs. (3.22) and (3.24) for some large polyatomic structures, involving dozens of atoms (azoles, fullerenes, nitroanilines, hydrocarbons), the data for which were not used in obtaining the approximations given by Eqs. (3.22) and (3.24), with the available reference data [34, 47, 88, 115–119] (see Fig. 3.20).

Considering the large experimental uncertainties (up to a factor 5–6) and marked scatter of theoretical estimates shown in Fig. 3.20, one may recognize that the obtained expressions yield reliable values of $\Delta\alpha^{ZPV}$ and $\Delta\alpha^{PV}$ corrections. For this reason, they can be employed to evaluate vibrational contributions to polarizability not only with the low computational burden (mainly due to the cost of the harmonic vibrational analysis) but also with an accuracy comparable with uncertainties of experiment and regular theoretical calculations.

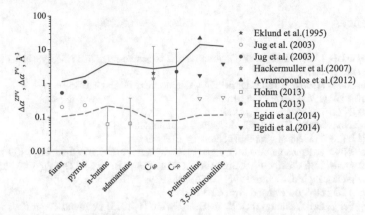

Fig. 3.20 The values of $\Delta\alpha^{\text{ZPV}}$ (dashed line, open symbols) and $\Delta\alpha^{\text{PV}}$ (solid line, closed symbols) for large polyatomic structures, found from Eqs. (3.22) and (3.24), as compared to the known data for furan and pyrrole (estimates [88]), n-butane and adamantane (differences between the RI measurements [47] and α_{el} obtained at the UB97-2/Sadlej computational level), C_{60} (measurements [115] and difference between RI measurements [116] and α_{el} calculations [118]), C_{70} (difference between the DP measurements [47] and α_{el} calculations [117]), p-nitroaniline and 3,5-dinitroaniline (estimates [34, 119]). Reported experimental uncertainties are shown by error bars (Reused with permission from Ref. [41]. Copyright 2017 IOP Publishing Ltd)

Finally, it should be stressed that for the structures involving 10–100 atoms, the pure vibrational contribution to observable polarizability can be as high as $\sim 10\ \text{Å}^3$ (amounting to 10–50% of the overall polarizability). In this regard, it would be fruitful to inspect how the $\Delta\alpha^{\text{ZPV}}$ and $\Delta\alpha^{\text{PV}}$ values change with regular increasing particle size. For simplicity, we consider a series of monoelement atomic clusters of variable size N. Primarily, as the particle size enlarges, the sum $\sum_i \frac{1}{\theta_{\text{vib},i}}$ in Eqs. (3.22) and (3.24) goes up with the number of modes $(3N - 6)$ that lead to the $\Delta\alpha^{\text{ZPV}}$ and $\Delta\alpha^{\text{PV}}$ growth (this also can be seen from Table 3.3 by the example of molecules of different sizes). However, this effect is further compensated by the decrease in the reduced rotational temperature $\theta_{\text{rot}}^{\text{red}}$ (due to the increment in the moment of inertia). As for the large-N asymptotic behavior, we can obtain, implying that $\alpha_{\text{el}} \propto N$, $R_{\text{CM}} \propto N^{1/3}$, $\theta_{\text{rot}}^{\text{red}} \propto N^{-5/3}$, $\sum_i \frac{1}{\theta_{\text{vib},i}} \propto N$ (see [120] for details of scaling), that for the $p_0..p_7$ magnitudes listed above, the $\Delta\alpha^{\text{ZPV}}(N)$ and $\Delta\alpha^{\text{PV}}(N)$ dependences scale as $\mathcal{O}(1)$ and $\mathcal{O}(N^{0.6})$, respectively.

Interestingly enough, the near constancy of the $\Delta\alpha^{\text{ZPV}}$ and $\Delta\alpha^{\text{PV}}$ corrections for large enough atomic clusters implied by the model [41] is experimentally consistent. Indeed, the measured $\Delta\alpha^{\text{PV}}$ magnitudes for C_{60} ($2.0\ \text{Å}^3$ [115]) and C_{70} ($2.21\ \text{Å}^3$ [41]) fullerenes prove to be only slightly higher than that predicted for the C_5 cluster ($0.61\ \text{Å}^3$). Besides, the ratio between the $\Delta\alpha^{\text{PV}}$ corrections for C_{60} and C_{70} matches the $N^{0.6}$ scaling law almost exactly.

References

1. Landau LD, Lifshitz EM (1977) Quantum mechanics: non-relativistic theory, vol 3. Pergamon Press, New York
2. Numerov B (1927) Astron Nachrichten (Astronomical Notes) 230:359
3. Cooley JW (1961) Math Comp 15:363
4. Bass JN (1972) J Comput Phys 9:555
5. Kurtz HA, Stewart JJ, Dieter KM (1990) J Comput Chem 11:82
6. Piela L (2007) Ideas of quantum chemistry. Elsevier
7. Ingamells VE, Papadopoulos MG, Handy NC, Willetts A (1998) J Chem Phys 109:1845
8. Loukhovitski BI, Sharipov AS, Starik AM (2016) J Phys B: At Mol Opt Phys 49:125102
9. Bishop DM, Pipin J, Silverman JN (1986) Mol Phys 59:165
10. Bishop DM (1990) Rev Mod Phys 62:343
11. Peng-Fei L, Lei Y, Zhong-Yuan Y, Yu-Feng G, Tao G (2013) Commun Theor Phys 59:193
12. Bytautas L, Matsunaga N, Ruedenberg K (2010) J Chem Phys 132:074307 (15 pp)
13. Coxon JA, Hajigeorgiou PG (2004) J Chem Phys 121:2992
14. Komasa J, Piszczatowski K, Lach G, Przybytek M, Jeziorski B, Pachucki K (2011) J Chem Theory Comput 7:3105
15. Gurvich LV, Veyts IV, Alcock CB (1989) Thermodynamics properties of individual substances. Hemisphere Pub. Co., New York
16. Loukhovitski BI, Starik AM (2009) Chem Phys 360:18
17. Lodi L, Tennyson J (2010) J Phys B: At Mol Opt Phys 43:133001
18. Szalay PG, Holka F, Fremont J, Rey M, Peterson KA, Tyuterev VG (2011) Phys Chem Chem Phys 13:3654
19. Pazyuk EA, Pupyshev VI, Zaitsevskii AV, Stolyarov AV (2019) Russ J Phys Chem A 93:1865
20. Billingsley FP II (1975) J Chem Phys 62:864
21. Sileo RN, Cool T (1976) J Chem Phys 65:117
22. Zemke WT, Stwalley WC, Langhoff SR, Valderrama GL, Berry MJ (1991) J Chem Phys 95:7846
23. Goorvitch D (1994) Astrophys J Suppl 95:535
24. Langhoff S, Bauschlicher C Jr (1995) J Chem Phys 102:5220
25. Maroulis G (1996) J Phys Chem 100:13466
26. Piecuch P, Spirko V, Kondo AE, Paldus J (1998) Mol Phys 94:55
27. Maroulis G (2003) J Mol Struct (THEOCHEM) 633:177
28. Kolos W, Wolniewicz L (1967) J Chem Phys 46:1426
29. Rychlewski J (1980) Mol Phys 41:833
30. Dyer MJ, Bischel WK (1991) Phys Rev A 44:3138
31. Marti J, Andres JL, Bertran J, Duran M (1993) Mol Phys 80:625
32. Bishop DM, Norman P (1999) J Chem Phys 111:3042
33. Helgaker T, Coriani S, Jorgensen P, Kristensen K, Olsen J, Ruud K (2012) Chem Rev 112:543
34. Egidi F, Giovannini T, Piccardo M, Bloino J, Cappelli C, Barone V (2014) J Chem Theory Comput 10:2456
35. Christiansen O, Hattig C, Gauss J (1998) J Chem Phys 109:4745
36. Maroulis G (2003) J Chem Phys 118:2673
37. Medved' M, Urban M, Kellö V, Diercksen GH (2001) J Mol Struct (THEOCHEM) 547:219
38. Sharipov AS, Loukhovitski BI, Pelevkin AV, Kobtsev VD, Kozlov DN (2019) J Phys B: At Mol Opt Phys 52:045101
39. Ruud K, Astrand PO, Taylor PR (2000) J Chem Phys 112:2668
40. Russell AJ, Spackman MA (1997) Mol Phys 90:251
41. Sharipov AS, Loukhovitski BI, Starik AM (2017) J Phys B: At Mol Opt Phys 50:165101(19pp)
42. Russell AJ, Spackman MA (1995) Mol Phys 84:1239
43. Illinger KH, Smyth CP (1960) J Chem Phys 32:787
44. Bishop DM, Cheung LM (1982) J Phys Chem Ref Data 11:119

45. Ruud K, Jonsson D, Taylor PR (2000) Phys Chem Chem Phys 2:2161
46. Naves ES, Castro MA, Fonseca TL (2011) J Chem Phys 134:054315
47. Hohm U (2013) J Mol Struct 1054–1055:282
48. Raghavachari K, Pople JA (1978) Int J Quantum Chem 14:91
49. Maroulis G, Thakkar AJ (1990) J Chem Phys 93:4164
50. Maroulis G (1991) J Chem Phys 94:1182
51. Voisin C, Cartier A, Rivail JL (1992) J Phys Chem 96:1966
52. Tennyson J, Bernath PF, Brown LR, Campargue A, Csaszar AG, Daumont L, Gamache , Hodges JT, Naumenko OV, Polyansky OL, Rothman LS, Vandaele AC, Zobov NF, Al Derzia AR, Fabrie C, Fazliev AZ, Furtenbacher T, Gordon IE, Lodi L, Mizus II (2013) J Quant Spectrosc Radiat Transfer 117:29
53. Fowler PW, Raynes WT (1981) Mol Phys 43:65
54. Thomsen B, Yagi K, Christiansen O (2014) Chem Phys Lett 610–611:288
55. Polyansky OL, Ovsyannikov RI, Kyuberis AA, Lodi L, Tennyson J, Zobov NF (2013) J Phys Chem A 117:9633
56. Saito S, Matsumura C (1980) J Mol Spectrosc 80:34
57. Kuczkowski RL, Suenram RD, Lovas FJ (1981) J Am Chem Soc 103:2561
58. Ebenstein WL, Muenter JS (1984) J Chem Phys 80:3989
59. Lide DR, Haynes WM (eds) (2010) CRC handbook of chemistry and physics, 90th edn., vol 9, chap. Dipole moments. CRC Press, Boca Raton, Florida, pp 50–58
60. Hickey AL, Rowley CN (2014) J Phys Chem A 118:3678
61. Verma P, Truhlar DG (2017) Phys Chem Chem Phys 19:12898
62. Halkier A, Klopper W, Helgaker T, Jorgensen P (1999) J Chem Phys 111:4424
63. Sadlej AJ (1988) Collec. Czech. Chem Commun 53:1995
64. Sadlej AJ (1991) Theor Chim Acta 81:45
65. Rappoport D, Furche F (2010) J Chem Phys 133:134105
66. Mohajeri A, Alipour M (2012) J Chem Phys 136:124111
67. Sun H, Autschbach J (2013) ChemPhysChem 14:2450
68. Maroulis G (2013) Rep Theor Chem 2:1
69. Alipour M, Fallahzadeh P (2017) Theor Chem Acc 136:22
70. Maroulis G, Xenides D (1999) J Phys Chem A 103:4590
71. Bak KL, Gauss J, Helgaker T, Jørgensen P, Olsen J (2000) Chem Phys Lett 319:563
72. Hobson SL, Valeev EF, Csaszar AG, Stanton JF (2009) Mol Phys 107:1153
73. Arapiraca AFC, Jonsson D, Mohallem JR (2011) J Chem Phys 135:244313
74. Johns JWC, McKellar ARW (1978) Can J Phys 56:736
75. Riley G, Raynes WT, Fowler PW (1979) Mol Phys 38:877
76. Shostak SL, Ebenstein WL, Muenter JS (1991) J Chem Phys 94:5875
77. Callegari A, Theule P, Muenter JS, Tolchenov RN, Zobov NF, Polyansky OL, Tennyson J, Rizzo TR (2002) Science 297:993
78. Johns JWC, McKellar ARW (1975) J Chem Phys 63:1682
79. Vaccaro PH, Kinsey JL, Field RW, Dai H (1983) J Chem Phys 78:3659
80. Deleon RL, Muenter JS (1984) J Chem Phys 80:3992
81. Ozier I (1971) Phys Rev Lett 27:1329
82. Kagann RH, Ozier I, Gerry MCL (1976) J Chem Phys 64:3487
83. Ishibashi C, Sasada H (1997) J Mol Spectrosc 183:285
84. Mikhailov VM, Smirnov MA (2001) Opt Spectrosc 90:27
85. Arapiraca AFC, Mohallem JR (2016) J Chem Phys 144:144301
86. Uzer T, Miller WH (1991) Phys Rep 199:73
87. Šolc M, Herman Z (1992) Collect Czech Chem Commun 57:1157
88. Jug K, Chiodo S, Calaminici P, Avramopoulos A, Papadopoulos MG (2003) J Phys Chem A 107:4172
89. Pandey PKK, Santry DP (1980) J Chem Phys 73:2899
90. Pederson MR, Baruah T, Allen PB, Schmidt C (2005) J Chem Theory Comput 1:590
91. Torrent-Sucarrat M, Luis JM, Kirtman B (2005) J Chem Phys 122:204108

92. Lao KU, Jia J, Maitra R, DiStasio RA Jr (2018) J Chem Phys 149:204303
93. Bulanin MO, Burtsev AP, Tret'yakov PY, Khim Fiz (1988) Sov J Chem Phys 7:1615
94. Bishop DM, Kirtman B (1991) J Chem Phys 95:2646
95. Santiago E, Castro MA, Fonseca TL, Mukherjee PK (2008) J Chem Phys 128:064310
96. Raynes WT, Lazzeretti P, Zanasi R (1988) Mol Phys 64:1061
97. Kongsted J, Christiansen O (2007) J Chem Phys 127:154315
98. Ingamells VE, Papadopoulos MG, Raptis SG (1999) Chem Phys Lett 307:484
99. Russell AJ, Spackman MA (2000) Mol Phys 98:855
100. Russell AJ, Spackman MA (2000) Mol Phys 98:867
101. Miliordos E, Xantheas SS (2014) J Am Chem Soc 136:2808
102. Maroulis G (1994) J Chem Phys 101:4949
103. Phibbs MK, Giguere PA (1951) Can J Chem 29:173
104. Antoine R, Rayane D, Allouche AR, Aubert-Frecon M, Benichou E, Dalby FW, Dugourd P, Broyer M (1999) J Chem Phys 110:5568
105. Miller TM (2010) CRC handbook of chemistry and physics, 90th edn., vol 10, chap. Atomic and molecular polarizabilities. CRC Press, Boca Raton, Florida, pp 193–202
106. Osipov AI, Uvarov AV (1992) Sov Phys Usp 35:903
107. Aidas K, Angeli C, Bak KL et al (2014) WIREs Comput Mol Sci 4:269
108. Werner HJ, Meyer W (1976) Mol Phys 31:855
109. Xenides D, Maroulis G (2006) J Phys B: At Mol Opt Phys 39:3629
110. Li ZH, Truhlar DG (2008) J Am Chem Soc 130:12698
111. Adamowicz L (1988) J Chem Phys 89:6305
112. Maroulis G, Makris C (1997) Mol Phys 91:333
113. Maroulis G (2000) Chem Phys Lett 318:181
114. Avramopoulos A, Papadopoulos MG (2002) Mol Phys 100:821
115. Eklund PC, Rao AM, Wang Y, Zhou KA, Wang P, Holden JM, Dresselhaus MS, Dresselhaus G (1995) Thin Solid Films 257:211
116. Hackermuller L, Hornberger K, Gerlich S, Gring M, Ulbricht H, Arndt M (2007) Appl Phys B 89:469
117. Zope RR (2007) J Phys B: At Mol Opt Phys 40:3491
118. Kowalski K, Hammond JR, de Jong WA, Sadlej AJ (2008) J Chem Phys 129:226101
119. Avramopoulos A, Reis H, Papadopoulos MG, Conf AIP (2012) Proceedings 1504:616
120. Sharipov AS, Loukhovitski BI (2019) Struct Chem 30:2057

Chapter 4
Influence of Vibrational and Rotational Degrees of Freedom of Molecules on Their Optical Properties

As was shown in a previous Chapter, the excitation of vibrational and rotational states of molecules substantially affects dipole moment and static polarizability. The state-specific values of these electric properties for a molecule with given vibrational and rotational quantum numbers were reported for a set of considered species (both di- and polyatomics). Afterward, these data can be utilized to evaluate the influence of the excitation of internal degrees of freedom of molecules on the refractive index of gaseous media (as well as on their dielectric permittivity).

As is recognized, comprehensive data on the refractive index of molecular gases are needed for the non-invasive optical diagnostics of gasdynamic flows (shadowgraph, schlieren, and interferometry visualization techniques) at high temperature and when the thermodynamic equilibrium between internal and translational degrees of freedom is not reached (in active media of molecular lasers, behind strong shock waves, in electric discharge plasma, in expanding supersonic jets, etc.) [1–5]. The alteration in the refractive index of gas upon laser excitation of molecules can govern nonlinear effects related to the propagation of laser beams in gas and plasma (principally, beam self-focusing) [6–8]. Knowledge of refractivity of nonequilibrium air plasma also can help ensure the reliable operation of onboard instrumentation in hypersonic flight conditions [5, 9].

In the present Chapter, we will consider this issue from a methodological point of view, reveal the influence of vibrational and, if possible, rotational excitation on the optical properties of various gases, and also compare the numerical estimates of the corresponding effects with known experimental data for high-temperature gases.

A. S. Sharipov et al., *Influence of Internal Degrees of Freedom on Electric and Related Molecular Properties*, SpringerBriefs in Electrical and Magnetic Properties of Atoms, Molecules, and Clusters, https://doi.org/10.1007/978-3-030-84632-9_4

4.1 Diatomic Molecules

At low temperature, refractive index n (together with the dielectric constant ε) of molecular gas is mainly specified by isotropic polarizability $\alpha_{V,J}$ and dipole moment $\mu_{V,J}$ of molecules in low vibrational and rotational states [10]. However, at high temperatures as well as in the case of nonequilibrium excitation (or delayed excitation) of vibrational and rotational degrees of freedom, the dependence of $\alpha_{V,J}$ and $\mu_{V,J}$ on the vibrational V and rotational J quantum numbers comes into play and can contribute appreciably to the magnitude of n or ε (for nonmagnetic media, $\varepsilon = n^2$).

As generally recognized, the refractive index n of rarefied gas ($n \approx 1$) is determined via the gas density ρ and the Gladstone–Dale constant K^{GD} by the approximate relationship [3, 8]

$$n - 1 = \rho K^{GD} . \tag{4.1}$$

Note that for a dense gas, in order to relate the refractivity and the density it is preferable to use the exact Lorentz–Lorenz formula [6, 8]

$$\frac{n^2 - 1}{n^2 + 2} = \frac{2}{3} \rho K^{GD} . \tag{4.2}$$

The Gladstone–Dale "constant", in turn, is governed by [11, 12]

$$K^{GD} = \frac{1}{2} \frac{N_A}{M_G \varepsilon_0} \left(\sum_{V,J} 4\pi \varepsilon_0 \alpha_{V,J} N_{V,J} + \sum_V \frac{\mu_{V,J=0}^2}{3hc_0 B_e} N_{V,J=0} \right) . \tag{4.3}$$

Here h is the Planck constant, c_0 is the speed of light in vacuum, B_e is the rotational constant of a diatomic molecule, ε_0 is the vacuum dielectric permittivity, N_A is the Avogadro number, M_G is the molar mass of gas, $N_{V,J}$ are the relative populations of molecules in the given state specified by V and J quantum numbers. The second term in Eq. (4.3), the orientation part of the Gladstone–Dale constant, implies that only the ground rotational states ($J = 0$) of polar molecules can contribute to the K^{GD} magnitude.

Under nonequilibrium conditions, when there are local Boltzmann distributions over rotational and vibrational states with their own rotational T_{rot} and vibrational T_{vib} temperatures, and $T_{\text{rot}} \neq T_{\text{vib}}$, the relative population of vibrational-rotational level (V, J) can be stated as

$$N_{V,J} = \frac{q_{V,J}(2J + 1) \exp\left(-\frac{E_{V,0}}{k_b T_{\text{vib}}} - \frac{E_{V,J} - E_{V,0}}{k_b T_{\text{rot}}}\right)}{\sum_{V^*,J^*} q_{V^*,J^*}(2J^* + 1) \exp\left(-\frac{E_{V^*,0}}{k_b T_{\text{vib}}} - \frac{E_{V^*,J^*} - E_{V^*,0}}{k_b T_{\text{rot}}}\right)} , \tag{4.4}$$

where $q_{V,J}$ is the nuclear spin degeneracy, $E_{V,J}$ is the energy of a given vibrational-rotational level, k_b is the Boltzmann constant. When vibrational, rotational, and translational degrees of freedom of molecules are thermodynamically equilibrated, we have $T_{vib} = T_{rot} = T$, where T is the gas (translational) temperature.

First, let us examine a gas composed only of homonuclear diatomic molecules possessing zero permanent dipole moment. In this case, according to Eq. (4.3), the refractive index is governed only by the polarizability averaged over all vibrational and rotational states $\alpha(T)$. The $\alpha(T)$ dependences for H_2, N_2, and O_2 molecules obtained by averaging in line with Eq. (4.4) the $\alpha_{V,J}^{UP}$ values described in Sect. 3.1 for the case of thermal equilibrium gas ($T_{vib} = T_{rot} = T$) in [12] are presented in Fig. 4.1.

It can be seen that thermally equilibrium heating up the gas to a temperature of 2000 K leads to a certain increase in the polarizability of molecules averaged over the Boltzmann ensemble: by 2%, 0.9%, and 0.4% for H_2, O_2, and N_2 molecules, respectively. For comparison, the temperature dependences of averaged polarizability for H_2, N_2, and O_2 molecules determined previously both experimentally [13–17] and theoretically [4, 18, 19] are also given in Fig. 4.1. Here we observe adequate agreement between our estimates [12] and the findings of other researchers.

Notice that the experimental dependences $\alpha(T)$ for the H_2, N_2, and O_2 diatomic molecules [13–15] were built based on the precision measurements of the refractive index at wavelength $\lambda = 633$ nm, and, as a matter of fact, the immediate comparison of the calculated static polarizability values with these measurements is not entirely correct. Importantly, the numerical evaluation with the help of time-dependent Hartree-Fock approximation [20] exhibited that the frequency-dependent (dynamic) part of polarizability for the ground vibrational states of H_2, N_2, and O_2 molecules at $\lambda=633$ nm did not surpass 3% of total polarizability. Besides, it also can be considered independent of the vibrational and rotational quantum numbers [19, 21]. Hence, we are led to infer that such a comparison can be still quite representative.

The influence of exciting the molecular vibrations and rotations of H_2, N_2, and O_2 molecules on their averaged polarizability is displayed in Fig. 4.2. We see that in the case of strong vibrational excitation ($T_{vib} = 6000$ K), when the rotational temperature is low and equals translational one ($T_{rot} = T = 300$ K), the increment of the averaged polarizability amounts a factor of 1.01-1.06. In the case, when the rotational temperature is as high as the vibrational ($T_{rot} = T_{vib}$), the averaged polarizability increases more dramatically: the rise of the averaged polarizability is about two time greater than that in the case of exciting the molecular vibrations only. These dependences demonstrate that the simultaneous excitation of vibrational and rotational degrees of freedom of considered molecules can notably contribute to the enlargement of molecular polarizability, and the effect is most pronounced for molecular hydrogen.

We regret to note that the dependences depicted in Fig. 4.2 for a broad temperature range up to 6000 K cannot be directly confronted with the experimental refractivity data, since the present analysis does not allow for the influence of excited electronic states on the electric properties of molecules, but such a contribution can be substantial at high temperatures, so long as the electronically excited molecules can have greater polarizabilities compared to their ground-state counterparts [22, 23] (see also Chap. 5). The dissociation and ionization of molecules and atoms, causing

Fig. 4.1 Temperature dependence of normalized averaged polarizability $\alpha(T)/\alpha(T = 300\,\text{K})$ for H_2, N_2, and O_2 calculated in [12] together with the available experimental [13–17] and theoretical [4, 18, 19] data

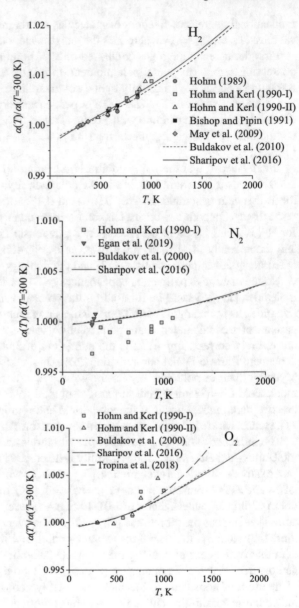

the formation of open-shell species and charged particles with their specific effect on refractivity, also contribute at high temperatures [1, 24–26] (see Chap. 6 for details).

Let us next analyze the gas of diatomic molecules possessing nonzero permanent dipole moment. The refractive index of such a gas depends on both the polarizability and the dipole moment of molecules, and, accordingly, separate averaging of $\alpha_{V,J}$ (a fortiori, $\mu_{V,J}$) over the vibrational-rotational states is somewhat meaningless, and

Fig. 4.2 Dependence of normalized averaged polarizability for the H_2, N_2, and O_2 molecules on vibrational temperature T_{vib} calculated for two cases: $T_{rot} = 300$ K (dashed curves) and $T_{rot} = T_{vib}$ (solid curves) (Reused with permission from Ref. [12]. Copyright 2016 IOP Publishing Ltd)

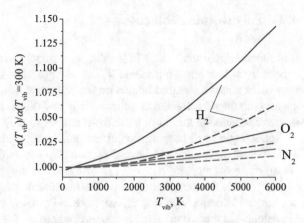

Fig. 4.3 The dependence of normalized Gladstone–Dale constant on vibrational temperature T_{vib} calculated at $T_{rot} = 300$ K for some polar molecules (Reused with permission from Ref. [12]. Copyright 2016 IOP Publishing Ltd)

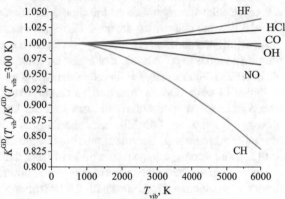

it makes real sense to average exactly the K^{GD} constant. Figure 4.3 represents the Gladstone–Dale constant averaged according to the distribution defined by Eq. (4.4) at $T_{rot} = 300$ K versus T_{vib} for CO, NO, OH, CH, HCl, and HF molecules. As seen, the highly diverse behavior of $K^{GD}(T_{vib})$ dependence for various gases occurs. This difference is caused mainly by the distinction in the $\mu_{V,J=0}$ values for molecules at issue. Specifically, as we saw in Sect. 3.1 for HCl and HF, the dependences of the dipole moment $\mu_{V,J=0}$ on vibrational quantum number V have distinct maxima at $V = 6$ and $V = 9$, respectively. This leads to an increase of K^{GD} values with the vibrational temperature rise. As for OH, CH, and NO species, their dipole moments diminish with V number, and the respective values of K^{GD} decrease with the vibrational temperature increment. The sharpest fall in the K^{GD} magnitude upon the excitation of vibrations is found for CH molecule. The resulting behavior of $K^{GD}(T_{vib})$ dependence for CO, in turn, is slightly nonmonotonic, so long as the dipole moment of CO changes its sign at $V = 3$. Hence, in light of the above, one can infer that when exciting molecular vibrations, the behavior of Gladstone–Dale constant (and, accordingly, of refractive index) with varying vibrational temperature can be very diverse for different molecular gases.

4.2 Polyatomic Molecules

It is supposed that the change in the dipole moment owing to vibrational excitation, especially at low gas temperature T and at high vibrational quantum numbers V, may bring about a sizeable impact on the refractive index of molecular polyatomic gases, both due to a change in polarizability and dipole moment, and this effect can be more substantial than for diatomic molecules [8, 27].

Let us begin by discussing the influence of the dipole moment. As outlined above, excitation of individual vibrational modes of polyatomic molecules can cause the change in dipole moment $\Delta\mu \sim 0.1$ D. To estimate the upper limits of a possible effect of such dipole moment alteration on the observable properties of molecular gases, one can perform the following calculations. As you know, the orientation part of the Gladstone–Dale constant K_{or}^{GD} is proportional to μ^2 (see Eq. 4.3). Taking $\mu = 1$ D as a representative magnitude of the dipole moment of polyatomic molecules, for $\Delta\mu = 0.1$ D one may anticipate a $\sim 20\%$ relative increment for K_{or}^{GD}.

As mentioned above in Sect. 3.2.2, the selective excitation of vibrational states of individual modes in polyatomic molecules is difficult to implement in practice due to fast intermode vibrational-vibrational exchanges at normal conditions. Such selective excitation of desired vibrational modes can be realized under specific nonequilibrium conditions and particularly for small molecules, for instance, behind the front of strong shock waves [28, 29], in nonequilibrium discharge plasma [30, 31], in a rarefied gas medium of the middle and upper atmosphere under the solar irradiation [32], and upon resonance absorption of laser radiation [33–36].

However in the thermal equilibrium case, vibrational levels of all the modes in the molecule are populated in accord with the Boltzmann distribution determined by the single gas temperature T, and the contributions of different modes in a total observed dipole moment can partially counteract each other (see, for instance, Figs. 3.13 and 3.14). Estimates indicate that the effect of temperature variation on the averaged dipole moment, in this case, is really small since the change in dipole moment upon heating from 10 to 3000 K normally does not go beyond ± 0.05 D [37]. Hence, possible consequences of the alteration of the dipole moment with temperature variation are less marked than in the case of selective excitation of vibrations of certain modes.

Now we turn to the contribution to the optical properties of gases related to polarizability. Given the state-specific values of $\Delta\alpha_{m,V_m}^{UP}$ and $\Delta\alpha_{m,V_m}^{PP}$, one can calculate, according to Eq. (4.3), the overall contribution of vibrational motion to the polarizability of gas at the desired distribution of polyatomic molecules over vibrational levels. To this end, both state-to-state [29, 31] and mode approximations for thermal nonequilibrium gas flows [35, 38, 39] are to be applied to specify populations of vibrational levels of polyatomic molecules. As can be deduced from Fig. 3.17 and Fig. 3.18, the excitation of individual vibrational modes of polyatomic molecules up to the preselected limit E_{cut} can induce rather different magnitudes of $\Delta\alpha_{m,V_m}^{UP}$ and $\Delta\alpha_{m,V_m}^{PP}$ corrections (over the range of 1–20% of α_{el}).

However, for many practically important cases, the vibrational (generally, together with rotational) and translational degrees of freedom of molecules are in thermodynamic equilibrium, and the effect of vibrations on refractivity can be estimated for the Boltzmann distribution of molecules among the vibrational levels. In this regard, it would be interesting to confront the derived data on $\Delta\alpha^{UP}_{m,V_m}$ with the temperature dependences of polarizability, measured with precision interferometric techniques in different nonpolar gases and reported elsewhere [14, 21, 40, 41]. Since these data were determined for the case of thermal equilibrium, the $\Delta\alpha^{UP}_{m,V_m}$ values must be averaged over the Boltzmann distribution as follows:

$$\Delta\alpha^{UP}(T) = \sum_{m=1}^{N} \frac{\sum_{V=1}^{V_m^{max}} \Delta\alpha^{UP}_{m,V_m} \exp\left(-\frac{E_{m,V_m}-E_{m,0}}{k_b T}\right)}{\sum_{V=0}^{V_m^{max}} \exp\left(-\frac{E_{m,V_m}-E_{m,0}}{k_b T}\right)}, \tag{4.5}$$

where V_m^{max} is a number of the highest vibrational level of a mode m. Note also that the degeneracies of vibrational modes in Eq. (4.5) are implicitly set to unity because in the present book the degenerated modes are treated separately. The $\Delta\alpha^{PP}_{m,V_m}$ values can be averaged similarly.

The temperature dependences $\Delta\alpha^{UP}(T)$ and $\Delta\alpha^{PP}(T)$ for considered molecules were calculated in this manner and tabulated for the broad temperature range ($T = 10 - 3000\,\mathrm{K}$) in [37]. Following [21], these data can be approximated by the quadratic function with two fitting parameters b and c:

$$\Delta\alpha^{UP/PP}(T) = b^{UP/PP}T + c^{UP/PP}T^2. \tag{4.6}$$

Corresponding $b^{UP/PP}$ and $c^{UP/PP}$ values both for $\Delta\alpha^{UP}(T)$ and $\Delta\alpha^{PP}(T)$ quantities are summarized in Tables 3.4 and 3.5 of Chap. 3.

Figure 4.4 displays the temperature dependences of averaged static polarizability for CO_2 and CH_4 molecules, calculated as follows

$$\alpha(T) = \alpha^{MP4}_{el} + \Delta\alpha^{ZPV} + \Delta\alpha^{UP}(T). \tag{4.7}$$

The experimental data extracted from the precise refractivity measurements [14, 21] and theoretical estimates [42] are given here for comparison as well. Note that to derive static polarizability values, the findings of Hohm and Kerl [14], reported for wavelength $\lambda = 633$ nm, are corrected for the dynamic wavelength-dependent part of polarizability, calculated employing the time-dependent DFT methodology (the magnitude of the correction factor is 0.9813 [37]). Also, in Fig. 4.4 the temperature dependence reported by Kongsted and Christiansen [42] is shifted by 0.152 Å^3 downwards to match the low-temperature limit of Kerl et al. [21]. We see here that though the computational strategy [37] based on the MP4(SDQ) α_{el} values overestimates to a little degree the low-temperature values of CO_2 and CH_4 polarizability (by 1% and 0.3%, respectively), a superb agreement for $\Delta\alpha^{UP}(T)$ behavior between the calculations [37] and the data of other studies is found.

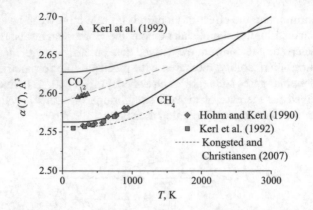

Fig. 4.4 Temperature dependence of averaged static polarizability for the CO_2 and CH_4 molecules calculated in [37] in UP approximation (solid curves) together with the available experimental [14, 21] (symbols, linear extrapolation of data [21] for CO_2 is shown by dashed line) and theoretical data of Kongsted and Christiansen [42] for CH_4 (dotted curve)

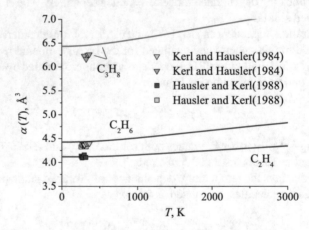

Fig. 4.5 Temperature dependence of averaged static polarizability for the C_2H_4, C_2H_6 and C_3H_8 molecules calculated in [37] in UP approximation (solid curves) together with the known experimental data [40, 41] (symbols) (Reused with permission from Ref. [37]. Copyright 2017 IOP Publishing Ltd)

Figure 4.5 display the comparison of the calculated temperature dependences of averaged polarizability for C_2H_4, C_2H_6, and C_3H_8 hydrocarbon molecules with the interferometric measurements [40, 41]. Notice, in this connection, that the experimental values, extracted from the refractive index at $\lambda = 633$ nm [41] and $\lambda = 546$ nm [40], were corrected here for dynamic part of polarizability as well (by the factors 0.9813 and 0.9750, respectively). Although the narrowness of the temperature ranges covered in measurements does not allow for a conclusive comparison of the temperature dependences [37] with experiments, we may observe for polariz-

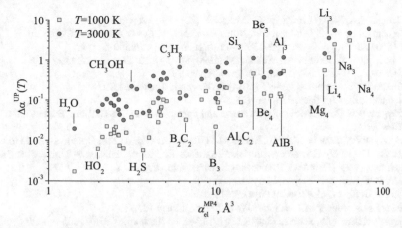

Fig. 4.6 The values of $\Delta\alpha^{UP}(T)$ correction at different temperatures for polyatomic species under consideration versus corresponding α_{el}^{MP4} values

ability at the temperature range $T = 250 - 370$ K a sound agreement between the calculations [37] and experiment.

To complement the picture, the values of $\Delta\alpha^{UP}(T)$ correction (note that the $\Delta\alpha^{ZPV}$ term is not included) for polyatomic species under consideration are presented in Fig. 4.6 against corresponding α_{el}^{MP4} values for two temperatures ($T = 1000$ K and $T = 3000$ K). An approximate proportionality between $\Delta\alpha^{UP}(T)$ and α_{el}^{MP4} contributions is observed. Eventually, we can conclude that for all the considered species the temperature effect on the polarizability is noticeable: the relative increase in polarizability within UP approximation (and, consequently, in refractive index) upon temperature rise till 3000 K typically lies within 3–4% of the electronic part. However for several clusters (Li$_3$, Li$_4$, Na$_3$, Si$_3$, Na$_4$, Mg$_4$, Al$_3$, Be$_3$, Be$_4$, Al$_2$C$_2$, AlB$_3$) this relative increase can reach 4%-5% of α_{el}^{MP4}. As for the $\Delta\alpha^{PP}(T)$ correction, the overall picture is the same as for the $\Delta\alpha^{UP}(T)$ term, and the relative increase in polarizability in PP approximation (and, therefore, in dielectric permittivity) upon temperature rise till 3000 K lies within 2–5% of the electronic part of polarizability.

Thereby, vibrational excitation of polyatomic species can lead to a significant change in refractive index (and, at the same time, dielectric permeability) of gas mixtures containing them when heated to 3000 K. We note, in passing, that the average polarizability of diatomic molecules, as a rule, increases only by approximately 2–3% with such a rise in temperature (see Sect. 4.1).

References

1. Kharitonov AI, Khoroshko KS, Shkadova VP (1974) Fluid Dyn 9:851
2. Yun-yun C, Zhen-hua L, Yang S, An-zhi H (2009) Appl Opt 48:2485
3. Wang M, Mani A, Gordeyev S (2012) Annu Rev Fluid Mech 44:299

4. Tropina AA, Wu Y, Limbach CM, Miles RB (2018) AIAA paper, pp 3904
5. Tropina AA, Wu Y, Limbach CM, Miles RB (2019) J Phys D Appl Phys 53:105201
6. Osipov AI, Filippov AA (1989) J Eng Phys Thermophys 56:590
7. Zhuravlev VV, Sorokin AA, Starik AM (1990) Sov J Quant Electron 20:435
8. Osipov AI, Uvarov AV (1992) Sov Phys Usp 35:903
9. Takahashi Y, Yamada K, Abe T (2014) J Spacecraft Rockets 51:1954
10. Hohm U, Kerl K (1986) Mol Phys 58:541
11. Flygare WH (1978) Molecular structure and dynamics. Prentice-Hall Inc, Englewood Cliffs, New Jersey
12. Sharipov AS, Loukhovitski BI, Starik AM (2016) J Phys B: At Mol Opt Phys 49:125103
13. Hohm U (1989) Hochtemperaturinterferometrie an gasen. Technische Universitat Braunschweig (West Germany), Braunschweig Dissertation thesis
14. Hohm U, Kerl K (1990) Mol Phys 69:803
15. Hohm U, Kerl K (1990) Mol Phys 69:819
16. May EF, Moldover MR, Schmidt JW (2009) Mol Phys 107:1577
17. Egan PF, Stone JA, Scherschligt JK (2019) J Vac Sci Technol, A 37:031603
18. Bishop DM, Pipin J (1991) Mol Phys 72:961
19. Buldakov MA, Matrosov II, Cherepanov VN (2000) Opt Spectrosc 89:37
20. Korambath P, Kurtz HA (1996). In: Karna SP, Yeates AT (eds) Nonlinear optical materials. Washington DC, pp 133–144
21. Kerl K, Hohm U, Varchmin H (1992) Ber Bunsenges Phys Chem 96:728
22. Urban M, Sadlej AJ (1990) Theor Chim Acta 78:189
23. Paleníková J, Kraus M, Neogrády P, Kellö V, Urban M (2008) Mol Phys 106:2333
24. Alpher RA, White DR (1959) Phys Fluids 2:153
25. Alpher RA, White DR (1959) Phys Fluids 2:162
26. Gladkov SM, Koroteev NI (1990) Sov Phys Usp 33:554
27. Osipov AI, Panchenko VY, Filippov AA (1984) Sov J Quantum Electron 14:1259
28. Armenise I (2017) Chem Phys 491:11
29. Armenise I, Kustova E (2018) J Phys Chem A 122:8709
30. Fridman A (2008) Plasma Chemistry. Cambridge University Press, Cambridge, UK
31. Kozak T, Bogaerts A (2014) Plasma Sources Sci Technol 23:045004
32. Solomon S, Kiehl JT, Kerridge BJ, Remsberg EE, Russell JM III (1986) J Geophys Res 91:9865
33. Lawrance WD, Knight AEW (1982) J Chem Phys 76:5637
34. Bloembergen N, Zewail AH (1984) J Phys Chem 88:5459
35. Starik AM, Titova NS, Loukhovitski BI (2004) Tech Phys 49(1):76
36. Lukhovitskii BI, Starik AM, Titova NS (2004) Kinet Catal 45(6):847
37. Sharipov AS, Loukhovitski BI, Starik AM (2017) J Phys B: At Mol Opt Phys 50:165101(19pp)
38. Starik AM, Lukhovitskii BI, Titova NS (2008) Combust Explos Shock Waves 44(3):249
39. Skrebkov OV, Karkach SP, Ivanova AN, Kostenko SS (2009) Kinet Catal 50(4):461
40. Kerl K, Hausler H (1984) Ber Bunsenges Phys Chem 88:992
41. Hausler H, Kerl K (1988) Int J Thermophys 9:117
42. Kongsted J, Christiansen O (2007) J Chem Phys 127:154315

Chapter 5
Polarizability of Electronically Excited States

Electric properties of molecules in excited electronic states are far less known and less frequently studied (both experimentally and theoretically) than their ground-state counterparts [1–3] despite the fact that these properties are of growing interest in different fields, e.g., in design of substances and materials with large nonlinear optical properties [4, 5] (particularly, they also play a key role in understanding the solvent effects on frequency shifts in electronic spectra [2, 6, 7]) and in the kinetics and transport of thermally nonequilibrium reacting gas flows [8–10]. The point is that the polarizabilities of molecules in the excited electronic states can be quite different from those in the ground states [11, 12], as a result of which such molecules will have special optical and transport properties. Specifically, electronic excitation of molecules upon absorption of intense laser radiation by a gaseous medium can lead to a laser beam self-focusing due to a change in the refractive index of molecular gas [13, 14]. The knowledge of polarizability of excited states is also vital for optical (mainly, laser) diagnostics of excited species in nonequilibrium gas flows [15, 16]. Besides, the excited state electric properties enter the models of the excited state intermolecular interaction and the models of ion-molecule reactions as the empirical parameters [17–19]. At last, basic electric properties like dipole moments and dipole polarizabilities of molecules in their excited electronic states provide very important information about the features of the electronic structure of these states [3, 20] and, accordingly, on their chemical reactivity [21–23].

Note that the theoretical evaluation of molecular electric properties in excited electronic states is a rather problematic task even for the modern quantum chemical machinery. The point is that only a few of the electronic structure methods are capable to treat electronic excitation. Excited electronic states require a departure from the single determinant reference assumption and often display large static and dynamic correlation effects [1, 2, 11, 24]. As a result, advanced quantum chemical methods, which take into account the multireference character and electron corre-

A. S. Sharipov et al., *Influence of Internal Degrees of Freedom on Electric and Related Molecular Properties*, SpringerBriefs in Electrical and Magnetic Properties of Atoms, Molecules, and Clusters, https://doi.org/10.1007/978-3-030-84632-9_5

lation effects, like the multiconfiguration self-consistent field (MCSCF) [25] or the equation-of-motion (EOM) CC [26, 27] methods, must be employed for the accurate determination of electric properties of such molecules. In the last decade, time-dependent DFT (TDDFT) approach has become the method of choice for evaluation of the electronic excitation energies, especially for larger molecules and clusters where advanced wave function-based methods cannot be applied [28–30], though the success of TDDFT in reliable determination of the molecular electric properties critically depends on the choice of the functional employed [3].

It should be mentioned that for many atoms and molecules the electronic excitation leads to an increase in their polarizability pronounced to a greater or lesser degree [1, 3, 11, 36, 38]. This enhancement in response electric properties of electronic structure is particularly due to the population of a more diffuse, higher molecular orbitals [3, 27]. To represent this effect quantitatively, we use the quasi-classical description valid for highly-excited atomic states. It is known that within this framework, the electronic excitation of a hydrogen-like atom to the nth electronic level specified by the principal quantum number n leads to an increase of its static polarizability asymptotically by $\sim n^6$ times [13, 39]. Also, the energy of the atom in the nth level according to the Bohr model scales as $\sim n^{-2}$. With this in mind, it is not difficult to deduce that the ratio of the polarizabilities of the atom in the electronically excited state with the energy $T_e \gg 0$ and in the ground state $\xi = \alpha(T_e)/\alpha(T_e = 0)$ is governed by the simple expression

$$\xi = \frac{1}{\left(1 - \frac{T_e}{\text{IP}}\right)^{\gamma}} \tag{5.1}$$

where IP is an ionization potential of an atom, and $\gamma = 3$. As is known, for the hydrogen atom (nuclear charge $Z = 1$) IP$= 0.5E_h$, where E_h is the atomic energy unit (27.211 eV). Thus, with an increase in T_e, the polarizability of such excited atom increases unrestrictedly, formally reaching infinity as the excitation energy approaches its ionization potential.

A relative increase in the polarizability of a hydrogen-like atom with $Z = 1$ is shown in Fig. 5.1. From this figure, you can see that such an approximate formula based on the quasi-classical considerations is in a good agreement with the results of the exact solution for the hydrogen atom [40]. However, for many-electron atoms, such an analytical scaling given by Eq. (5.1) is apparently not valid, and numerical quantum chemical calculations are required to determine the polarizability of excited states. In particular, the data on the polarizability of the valence states of some neutral atoms (C, N, O, Si, P, and S) calculated using the multiconfiguration second-order perturbation theory by Andersson and Sadlej [31] are also summarized in Fig. 5.1. As you can see, the polarizability of electronically excited many-electron atoms, at least, at $T_e < 5$ eV grows much weaker than for the hydrogen atom. Such behavior, in principle, could be explained by large IP values for these atoms, but for these atoms, they are indeed not much greater or even less than the IP for a hydrogen atom (8.2–14.5 eV [41]). The true reason for the discrepancy between the predictions of the

Fig. 5.1 The relative polarizability ξ of quasi-classical excited H atom (IP$= 0.5E_h$) given by Eq. (5.1) (solid curve). The strict analytical calculations for H atom [40] and accurate quantum chemical calculations for different many-electron atoms [31] are also presented (symbols)

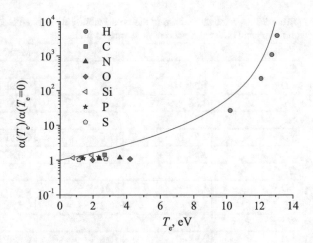

Fig. 5.2 The relative polarizability ξ of quasi-classical excited H atom (IP$= 0.5E_h$) given by Eq. (5.1) (solid curve). The accurate quantum chemical calculations for different molecules [7, 11, 16, 32, 33, 35, 37] are also presented (symbols)

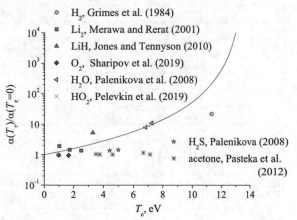

quasi-classical model and ab initio calculations, apparently, lies in the fundamental impossibility of describing response properties for complex electronic structures by simple analytic models.

As for molecules, the eventual effect of electronic excitation on the polarizability can be different even in sign. Figure 5.2 shows the ξ values for the electronically excited states of a number of molecules adopted from the literature [7, 11, 16, 32, 33, 35, 37]. The data on atoms and molecules used in the plotting of Figs. 5.1 and 5.2, as well as the ξ values for other molecules found in the literature, are shown in Tables 5.1 and 5.2. Note that the adiabatic (not vertical) polarizability values were selected for presentation and future analysis. As seen from Fig. 5.2 and Tables 5.1 and 5.2, the polarizability upon excitation, to one degree or another, increases for the most species. In particular, for some molecules (H$_2$, LiH, NaH, H$_2$O), the ξ values approximately coincide and even exceed the corresponding values for the hydrogen

Table 5.1 A relative increase in the polarizability for excited electronic states of different atoms and inorganic molecules according to various literature

State	T_e, eV	ξ	Method	Refs.
C				
^1D	1.26	1.15	CASPT2	[31]
^1S	2.68	1.43	CASPT2	[31]
N				
^2D	2.38	1.10	CASPT2	[31]
^2P	3.58	1.18	CASPT2	[31]
O				
^1D	1.97	1.01	CASPT2	[31]
^1S	4.19	1.08	CASPT2	[31]
Si				
^1D	0.78	1.16	CASPT2	[31]
^1S	1.91	1.40	CASPT2	[31]
P				
^2D	1.41	1.13	CASPT2	[31]
^2P	2.32	1.21	CASPT2	[31]
S				
^1D	1.15	1.03	CASPT2	[31]
^1S	2.75	1.09	CASPT2	[31]
H_2				
$B^1\Sigma_u^+$	11.34	21.91	MCSCF	[32]
Li_2				
$(1)^3\Sigma_u^+$	1.02	1.93	MCSCF	[33]
$(1)^3\Sigma_g^+$	1.73	1.46	MCSCF	[33]
$(2)^1\Sigma_g^+$	2.51	1.37	MCSCF	[33]
LiH				
$A^1\Sigma^+$	3.29	4.74	TDGI	[34]
		5.29	RMPS	[35]
NaH				
$A^1\Sigma^+$	2.82	2.59	TDGI	[34]
O_2				
$a^1\Delta_g$	0.98	0.95	CASSCF	[20]
		0.96	review	[36]
		0.97	XMCQDPT2	[16]
$b^1\Sigma_g^+$	1.66	0.96	XMCQDPT2	[16]
		0.96	LIGs (experiment)	[16]

(continued)

Table 5.1 (continued)

State	T_e, eV	ξ	Method	Refs.
CO				
$a^3\Pi$	6.04	1.34	RMPS	[35]
	5.70	1.26	XMCQDPT2//d-aug-cc-pVQZ	This work
HO_2				
$^2A'$	0.97	0.99	XMCQDPT2	[37]
H_2O				
$^3B_1$	6.81	7.73	CASPT2	[11]
	6.83	7.95	CCSD(T)	[11]
$^1B_1$	7.29	10.27	CASPT2	[11]
	7.26	11.02	CCSD(T)	[11]
H_2S				
$^3B_1$	4.42	1.36	CASPT2	[11]
	4.45	1.35	CCSD(T)	[11]
$^1B_1$	5.00	1.44	CASPT2	[11]
	5.07	–	CCSD(T)	[11]

atom, while for other molecules the excitation effect is much weaker. At the same time, for O_2 and HO_2, the polarizability upon excitation even decreases slightly.

Notice that multireference ab initio calculations (by using the extended multiconfiguration quasi-degenerate second-order perturbation theory method XMCQDPT2 [42]) and the corrections to polarizabilities for the nuclear zero-point motion have been recently employed to determine the static polarizabilities for the ground triplet ($X^3\Sigma_g^-$) and two excited singlet ($a^1\Delta_g$ and $b^1\Sigma_g^+$) electronic states of molecular oxygen [16]. It was revealed that the excitation of O_2 molecule into the lowest singlet electronic states leads to a decrease in its polarizability: the polarizabilities of $O_2(a^1\Delta_g)$ and $O_2(b^1\Sigma_g^+)$ are smaller (by factors of 0.967 and 0.962, respectively) than that of $O_2(X^3\Sigma_g^-)$. This feature of singlet oxygen may be associated with the fact that, unlike for most other molecules, the excitation of oxygen molecules to the $a^1\Delta_g$ state does not involve the population of a more diffuse, higher energy orbital (the same is valid for the $b^1\Delta_g^+$ state as well) [20, 24, 36, 43]. This fact also correlates with the observable regularities in transition intensities for absorption bands of molecular oxygen [36]. The theoretical results for the $b^1\Delta_g^+$ state qualitatively consistent with Williams' supposition made back in the late 1980s [44] were also quantitatively confirmed in [16] experimentally employing the diffraction at laser-induced gratings. Apparently, something similar is observed for HO_2 molecule [37].

We see that for both atoms and molecules, the general law for the dependence of ξ on T_e is not self-evident. However, due to the limited availability of reliable data on the polarizability of excited molecules and owing to the complexity of the

Table 5.2 A relative increase in the polarizability for excited electronic states of different organic molecules according to various literature

State	T_e, eV	ξ	Method	Refs.
C_2H_6O (acetone)				
$^3n - \pi^*$	3.50	1.04	CASPT2	[7]
$^1n - \pi^*$	3.75	1.02	CASPT2	[7]
$^3\text{ß} - \pi^*$	4.61	1.01	CASPT2	[7]
$^1\pi - \pi^*$	6.67	1.15	CASPT2	[7]
$^3\sigma - \pi^*$	7.13	1.02	CASPT2	[7]
C_6H_5 (CN) (benzonitrile)				
1^1B	4.85	1.10	CCSD	[3]
2^1A	5.87	1.44	CCSD	[3]
$C_4H_4N_2$ (pyrimidine)				
1^1B_2	4.59	1.20	EOM-CCSD	[27]
$C_2H_2N_4$ (s-tetrazine)				
1^1B_{1u}	2.39	1.10	EOM-CCSD	[27]
$C_4H_4N_2O_2$ (uracil)				
$1^3A'$	3.87	1.09	EOM-CCSD	[27]
$1^3A''$	4.95	1.07	EOM-CCSD	[27]
$1^1A''$	5.22	1.09	EOM-CCSD	[27]
$2^1A'$	5.58	1.36	EOM-CCSD	[27]
$C_6H_6N_2O_2$ (p-nitroanilin)				
1^3A_1	3.41	1.37	EOM-CCSD	[27]
2^1A_1	4.62	1.37	EOM-CCSD	[27]

associated quantum chemical calculations, it would be desirable to have a simple analytical model that allows extrapolating data for a few known electronic states to the states of arbitrary excitation energy. In principle, Eq. (5.1) for the Bohr atom is suitable for this purpose, however, in this case, we are forced to assume that in general (for many-electron atoms and molecules) the scaling laws for the energy and polarizability of the nth electronic level are different from the laws for the Bohr atom. Hereby, the parameter γ of Eq. (5.1) (equal to three for the Bohr atom) can take essentially any value and can be used as the single fitting parameter of the model.

It is interesting to examine what γ values correspond to the data presented in Tables 5.1 and 5.2, if Eq. (5.1) is used to describe them. These values obtained during the fitting of the data of Tables 5.1 and 5.2 by Eq. (5.1) are given in Fig. 5.3 as a function of the number of electrons in the system N_{el}. For this purpose, the IP values borrowed from [41, 45–47] were used. We see that the γ parameter is fairly close to three for small electronic systems (e.g., LiH, H_2, H_2O), while with the increase of the number of electrons the effect of electronic excitation on the polarizability significantly decreases. So for large polyatomic molecules such as acetone and ben-

Fig. 5.3 The obtained γ values for the species presented in Tables 5.1 and 5.2 versus the number of electrons in the system N_{el} (symbols) as well as the fitted power function trend to these data represented by Eq. (5.2) (curve)

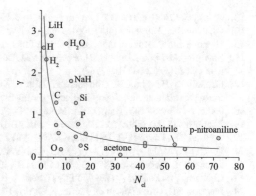

zonitrile, $\gamma < 0.5$. Approximately this behavior for γ can be described by a simple power-law as follows

$$\gamma(N_{el}) = 3.36 N_{el}^{-0.65}. \tag{5.2}$$

although, as can be seen from Fig. 5.3, the data scatter with respect to this approximation dependence is very significant. Perhaps this is due to the potential error in determining γ values from a small number of electronic states available for each molecule. Also it is not unlikely that the γ parameter is some way also depends on the other related quantities (for instance, IP, number of valence electrons, polarizability, dipole moment), but in order to trace such correlations it is necessary to collect a larger amount of initial data than it was done within the present Chapter. However, we hope that a detailed and comprehensive study of this issue based on novel theoretical and experimental data on the polarizability of molecules in excited electronic states will be carried out in the near future.

References

1. Urban M, Sadlej AJ (1990) Theor Chim Acta 78:189
2. Ruud K, Mennucci B, Cammi R, Frediani L (2004) J Comput Methods Sci Eng 4:381
3. Medved' M, Budzák Š, Pluta T (2015) Theor Chem Acc 134:78
4. Kanis DR, Ratner MA, Marks TJ (1994) Chem Rev 94:195
5. Bredas JL, Cornil J, Beljonne D, Santos DAD, Shuai Z (1999) Acc Chem Res 32:267
6. DeFusco A, Minezawa N, Slipchenko LV, Zahariev F, Gordon MS (2011) J Phys Chem Lett 2:2184
7. Pasteka LF, Melichercik M, Neogrady P, Urban M (2012) Mol Phys 110:2219
8. Kustova EV, Puzyreva LA (2009) Phys Rev E 80:046407
9. Pineda DI, Chen JY (2016) Western states section spring technical meeting of the combustion institute. Paper 139LF-0021
10. Kobtsev VD, Kostritsa SA, Smirnov VV, Titova NS, Torokhov SA (2020). Combust Sci Technol 192:744
11. Paleníková J, Kraus M, Neogrády P, Kellö V, Urban M (2008) Mol Phys 106:2333
12. Gupta K, Ghanty TK, Ghosh SK (2010) Phys Chem Chem Phys 12:2929

13. Askaryan GA (1966) JETP Lett (in Russian) 4:400
14. Gladkov SM, Koroteev NI (1990) Sov Phys Usp 33:554
15. Hemmerling B, Kozlov DN (2003) Chem Phys 291:213
16. Sharipov AS, Loukhovitski BI, Pelevkin AV, Kobtsev VD, Kozlov DN (2019) J Phys B: At Mol Opt Phys 52:045101
17. Capitelli M, Ferreira CM, Gordiets BF, Osipov AI (2000) Plasma kinetics in atmospheric gases. Springer series on atomic, optical, and plasma physics, vol 31. Springer, Berlin
18. Bultel A, Cheron BG, Bourdon A, Motapon O, Schneider IF (2006) Phys Plasmas 13:043502
19. Sharipov AS, Loukhovitski BI, Starik AM (2016) J Phys B: At Mol Opt Phys 49:125103
20. Poulsen TD, Ogilby PR, Mikkelsen KV (1998) J Phys Chem A 102:8970
21. Krech RH, McFadden DL (1977) J Am Chem Soc 99:8402
22. Ghanty TK, Ghosh SK (1993) J Phys Chem 97:4951
23. Chattaraj PK, Poddar A (1999) J Phys Chem A 103:1274
24. Paterson MJ, Christiansen O, Jensen F, Ogilby PR (2006) Photochem Photobiol 82:1136
25. Schmider HL, Becke AD (1998) J Chem Phys 108:9624
26. Stanton JF, Bartlett RJ (1993) J Chem Phys 98:7029
27. Nanda KD, Krylov AI (2016) J Chem Phys 145:204116
28. Bousquet D, Fukuda R, Maitarad P, Jacquemin D, Ciofini I, Adamo C, Ehara M (2013) J Chem Theory Comput 9:2368
29. Makowski M, Hanas M, Phys Z (2016) Chemistry 230:1425
30. Sharipov AS, Loukhovitski BI (2019) Struct Chem 30:2057
31. Andersson K, Sadlej AJ (1992) Phys Rev A 46:2356
32. Grimes RM, Dupuis M, Lester WA Jr (1984) Chem Phys Lett 28:247
33. Merawa M, Rerat M (2001) Eur Phys J D 17:329
34. Merawa M, Begue D, Dargelos A (2003) J Phys Chem A 107:9628
35. Jones M, Tennyson J (2010) J Phys B: At Mol Opt Phys 43:045101
36. Minaev BF (2007) Russ Chem Rev 76:1059
37. Pelevkin AV, Sharipov AS (2019) Plasma Chem Plasma Process 39:1533
38. Fuentealba P, Simon-Manso Y, Chattaraj PK (2000) J Phys Chem A 104:3185
39. Delone NB, Krainov VP, Shepelyanskii DL (1983) Sov Phys Usp 26:551
40. McDowell K (1976) J Chem Phys 65:2518
41. Lide DR (ed) (2010) CRC handbook of chemistry and physics, 90th edn. CRC Press
42. Granovsky AA (2011) J Chem Phys 134:214113
43. Minaev BF, Minaeva VA (2001) Phys Chem Chem Phys 3:720
44. Williams JH (1988) Chem Phys Lett 147:585
45. Afeefy HY, Liebman JF, Stein SE (2010) NIST chemistry WebBook, NIST standard reference database number 69. Chap. Neutral thermochemical data. National Institute of Standards and Technology, Gaithersburg MD, 20899. http://webbook.nist.gov
46. Johnson RD III, NIST, (2010) computational chemistry comparison and benchmark database, NIST standard reference database number 101 release, 15a edn
47. Danovich D, Apeloig Y (1991) J Chem Soc Perkin Trans 2(12):1865

Chapter 6
The Effect of Nonequilibrium in Internal Degrees of Freedom of Molecules on Their Physical Properties

In previous chapters, it was shown how the excitation of internal degrees of freedom of molecules (rotational, vibrational, and electronic) affects their basic electric properties (dipole moment and polarizability). This can cause the variation of some essential physical properties of molecular gases, such as refractive index, intermolecular potentials, transport coefficients, and even rate constants of chemical reactions. These effects are relevant for various fields of molecular physics, chemical kinetics, and physical gas dynamics. In the present chapter, we will consider the specific mechanisms of this influence.

6.1 Refractive Index

The effect of the excitation of molecular vibrations and rotations on the variation of the refractive index was considered in detail in Chap. 4. However, under real nonequilibrium conditions, often, along with vibrational and rotational molecular degrees of freedom, the electronic states are excited as well [1–5]. The influence of dissociation of molecules and ionization of atoms, leading to the formation of open-shell species and charged particles (both ions and electrons) with their specific effect on refractivity can also be significant [6–9].

Therefore, in order to interpret the experimental data on the optical/dielectric density of a nonequilibrium gaseous medium, which is important for the problems of optical diagnostics in nonequilibrium flows (shadowgraph, schlieren, and interferometry visualization techniques) [10–14], for analyzing the propagation of radio waves in low-temperature plasmas [14, 15] and intense laser radiation in gas mixtures [10, 16, 17], in general, the effect of all internal degrees of freedom are to be taken into account.

© The Author(s), under exclusive license to Springer Nature Switzerland AG 2022   75
A. S. Sharipov et al., *Influence of Internal Degrees of Freedom on Electric and Related Molecular Properties*, SpringerBriefs in Electrical and Magnetic Properties of Atoms, Molecules, and Clusters, https://doi.org/10.1007/978-3-030-84632-9_6

6.1.1 Polarizability of a Two-Level System

To show the effect of just electronic excitation on the observed polarizability of a gas, we consider for simplicity a model molecule with only two electronic levels: the ground level 0 with energy $E = 0$ and a degeneracy g_0, and an excited level 1 with energy $E = T_e$ and a degeneracy g_1. In accordance with Chap. 5, the ratio of the polarizability of the molecule in these levels ξ is governed by Eq. (5.1) and depends on the T_e/IP ratio, where IP is an ionization potential of a molecule. Then, in the case of thermodynamic equilibrium at a finite temperature T, the ratio of the polarizability averaged over both electronic states $\langle\alpha\rangle$ to the polarizability in the ground electronic state $\alpha(T_e = 0)$ will be determined by analytic expression

$$\frac{\langle\alpha\rangle(T)}{\alpha(T_e = 0)} = \frac{g_0 + g_1 \exp(-T_e/T)\xi(T_e/\text{IP})}{g_0 + g_1 \exp(-T_e/T)}, \tag{6.1}$$

where $\xi(T_e/\text{IP})$ is defined by Eq. (5.1). By setting, without loss of generality, g_0 and g_1 degeneracies to unity, and assuming that the typical value of the parameter γ of Eq. (5.1) is 1.0 (see Fig. 5.3), one can trace the effect of electronic excitation of a model two-level system on the observed polarizability at different temperature.

For this purpose, Fig. 6.1 shows the $\langle\alpha\rangle(T)/\alpha(T_e = 0)$ dependence calculated for different T_e/IP values. We see that the influence of electronic excitation is greater, the higher this state lies. So, the equilibrium electronic excitation of the state with $T_e = 0.8\text{IP}$ results in the one-and-a-half increase of the observable polarizability at $T = 0.5T_e$, whereas the possibility of excitation of less high-lying states, as can be seen, makes a significantly smaller contribution to $\langle\alpha\rangle(T)$.

Of course, the considered two-level system is only an unrealistic simplification of real situations, while in fact electronic excitation of the molecule is characterized by

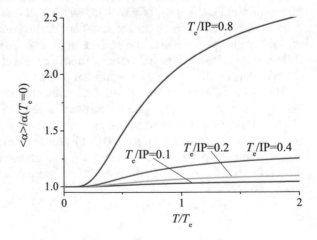

Fig. 6.1 The ratio of the averaged polarizability for a model two-level system with $g_0 = g_1 = 1$ calculated using Eqs. (6.1) and (5.1) for different T_e/IP ratios to the polarizability in the ground state against the reduced temperature T/T_e

the whole spectrum of electronic states. Besides, the excitation of electronic states is hardly separable from the excitation of other internal degrees of freedom, but the analytical approach used enables us to evaluate the order of the effect at issue.

6.1.2 Influence of Dissociation on Refractivity

As mentioned in the introduction to this Section, the refractive index of gas mixtures under substantially nonequilibrium conditions and at high temperatures is influenced to a notable extent by the dissociation of molecules. This occurs because the sum of the polarizabilities of the dissociation products is commonly greater than the polarizability of the source molecule. This is clearly seen in the example of diatomic molecules from Figs. 2.6, 2.7, 2.8, and 2.9 (Sect. 2.1), where the sum of the polarizability of atoms formed during the dissociation of the molecule significantly exceeds the polarizability of the molecule in the equilibrium nuclear configuration.

The point being that the polarizability of parts is greater than the polarizability of the whole can be interpreted both in terms of the Bader quantum theory of atoms in molecules [18] and using the well-known concept of pair polarizability [19] (see [20] for details). Meanwhile, this result is also a manifestation of the widely-used minimum polarizability principle (MPP) [21–23], according to which, "the natural direction of evolution of any molecular system is toward a state of minimum polarizability" (i.e. the polarizability of a molecule is inherently related to its thermodynamic and, probably, kinetic stability [21, 23–25]). So, in compliance with MPP, a molecule, being thermodynamically more stable than its constituent atoms, has a lower polarizability α_{mol} than the sum of the polarizability of atoms $\sum \alpha_{atom}$, i.e. the change in the static dipole polarizabilities upon dissociation (atomization) $\Delta\alpha_d = \sum \alpha_{atom} - \alpha_{mol} > 0$.

From general considerations, it is clear that the influence of dissociation (atomization) on the observed refractive index of the reacting gas will be higher for those molecules for which the $\Delta\alpha_d$ value is greater and the atomization energy (or enthalpy) is less. In this regard, it would be interesting to see, for simplicity, using only diatomic species as a test case, how the dissociation enthalpy ΔH_d^{298} can correlate with the increase in polarizability $\Delta\alpha_d$ during their decay.

The values for bond dissociation enthalpies ΔH_d^{298} and the change in the static dipole polarizabilities $\Delta\alpha_d$ for 19 diatomic species in their ground electronic states based on existing literature were used in the present analysis. In doing so, the thermochemical data were borrowed from [26] for molecules and from [27] for rare-gas dimers. The required polarizability values for atoms were adopted from the very recent review of Schwerdtfeger and Nagle [28]. The polarizability values for most diatomic molecules α_{mol} were taken from the exhaustive critical review of Hohm [29]. Due to the lack of data in the latter compilation for OH and CH radicals, we took the necessary values from [30] (1.14 $Å^3$) and from [31] (1.88 $Å^3$), respectively.

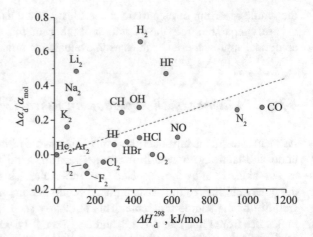

Fig. 6.2 The values of $\Delta\alpha_d / \alpha_{mol}$ versus ΔH_d^{298} for different diatomic species (symbols) and overall linear trend $\Delta\alpha_d / \alpha_{mol} = 3.7 \times 10^{-4} \, \Delta H_d^{298}$ (dotted line)

In Fig. 6.2 the relative change in the static polarizability in course of dissociation $\Delta\alpha_d / \alpha_{mol}$ for different diatomic species is shown in dependence of the ΔH_d^{298} enthalpy difference. We see that there is no unambiguous relationship between these two quantities, although there is a general trend indicating a positive correlation between $\Delta\alpha_d$ and ΔH_d^{298} quantities and therefore consistent with the MPP.

A direct proportionality trend with a positive slope describing the behavior for all 19 species is also depicted in Fig. 6.2. One can observe that in general, a greater contribution to the polarizability of the gas mixture during decomposition (per molecule) will come from those molecules that have a higher dissociation enthalpy, but, as a result, the dissociation of which is hampered. However, a number of molecules will give rise to a relative increase in polarizability upon dissociation significantly greater than for other molecules with approximately the same dissociation energies. They are H_2, Li_2, Na_2, and HF. At the same time, for individual species (viz. F_2, Cl_2, and I_2) it turned out that $\Delta\alpha_d < 0$ (i.e., these molecules, being thermodynamically more stable than its constituent atoms, have higher polarizabilities than the sums of the polarizability of atoms). That is, the dissociation of such molecules will lead to a decrease in the refractive index of the gas mixture. In fact, this can be considered a violation of the MPP, which, incidentally, happens not so rarely [32, 33]. This point should not surprise us since the MPP is not a rigorous law but it is only a useful and fruitful rule-of-thumb [33].

6.1.3   Refractivity of a Nonequilibrium Reacting Gas

We now consider how the dependence of the refractive index of the molecular gases on excitation of their internal degrees of freedom can affect the analysis of refractivity measurements (using, for example, the schlieren and shadowgraph imaging techniques) in nonequilibrium gas flows. To demonstrate these effects, molecular oxygen was chosen because it differs from other widespread diatomic gases by the presence of relatively low-lying electronic states [34], and, besides, the polarizabilities of the lower electronic states of O_2 ($X^3\Sigma_g^-$, $a^1\Delta_g$, and $b^1\Sigma_g^+$) has recently been carefully studied both theoretically and experimentally [35] (see Chap. 5). As an example of conditions in which strong thermal nonequilibrium is realized, we consider the flow behind a shock wave (SW) front.

Employing the thermal nonequilibrium kinetic model for reacting N_2/O_2 mixture developed by Kadochnikov and Arsentiev in [36] within the mode approximation formulated in [37], we obtained the spatial profiles of the flow parameters behind the steady SW front for the incoming flow Mach number $M_0 = 8$, the temperature and pressure in front of the SW $T_0 = 300$ K and $P_0 = 0.1$ bar. Under these boundary conditions, in line with the well-known Rankine–Hugoniot relations within the supposition that the gas chemical composition and vibrational temperatures remain unchanged across the SW front [2, 38], the gas temperature immediately behind the shock front is $T_1 = 4020$ K (the pressure $P_1 = 7.5$ bar).

To reveal the influence of various factors on refractivity, these flow parameters were chosen so that two conditions were satisfied: first, the temperature T_1 must be high enough to achieve substantial dissociation of O_2, and secondly, it should not exceed the temperature, after which the vibrational relaxation of O_2 and its dissociation start to occur on the same spatial scale (approximately, at $T_1 > 5000$ K).

Note that within a mode approximation for nonequilibrium flows, it is accepted that translational and rotational degrees of freedom of reacting molecules are equilibrated and within each vibrational mode there exists a local Boltzmann distribution with its own vibrational temperature. In doing so, nonequilibrium vibrational excitation was taken into account only for the ground electronic state $O_2(X^3\Sigma_g^-)$, because the concentrations of electronically excited states of O_2 and O_3 are expected to be small (their vibrational temperatures were assumed equal to translational).

During the calculations of the refractive index of reacting gas mixture, the state-specific polarizability values of $O_2(X^3\Sigma_g^-)$ reported in [31] were used. In addition, we adopted the values of polarizability for the other species abundant behind a SW front in oxygen flow from the literature ($O_2(a^1\Delta_g)$ [35], $O_2(b^1\Sigma_g^+)$ [35], $O(^3P)$ [39], $O(^1D)$ [39, 40], O_3 [41]).

Figure 6.3 displays the dependence of the gas temperature T, vibrational temperature T_{vib} of the $X^3\Sigma_g^-$ molecular oxygen, and the refractive index $n - 1$ on the distance behind the SW front in molecular oxygen. Here, we present both the total refractive index of the mixture and the contributions to refractivity from various nascent species originating during the dissociation of O_2 behind the SW. For convenience, the contributions from the products of O_2 dissociation are additionally zoomed in a separate subgraph.

We clearly see from Fig. 6.3a that, for the selected initial conditions, relaxation processes in O_2 proceed in two distinct stages. At the first stage (at $x < 0.05$ cm), $T_{\text{vib}} < T$ and vibrational relaxation occurs predominantly. Then (at $x > 0.05$ cm, this region can be designated as the dissociation stage), vibrations are equilibrated with translational and rotational degrees of freedom ($T_{\text{vib}} = T$), dissociation due to high

Fig. 6.3 The dependence of T and $T_{\text{vib}}(O_2(X^3\Sigma_g^-))$ (**a**) and the refractive index $n-1$ (**b**) on the distance behind the SW front x in molecular oxygen (incoming flow Mach number $M_0 = 8$, the temperature and pressure in front of the SW $T_0 = 300$ K and $P_0 = 0.1$ bar). Within graph (**b**) the solid lines show the full refractive index of the mixture ('full') and the contributions from the individual components of the mixture. The dashed curve stands for the refractive index of the mixture, obtained without allowance for the temperature dependence of the averaged polarizability for $O_2(X^3\Sigma_g^-)$

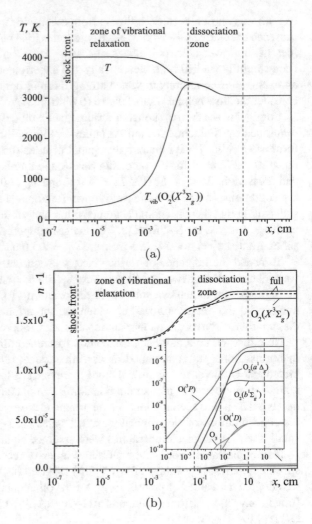

(a)

(b)

T_{vib} proceeds more intensively, and at distance about 5 cm from the SW front, the gas mixture comes into the complete thermodynamic equilibrium. Eventually, the gas temperature due to dissociation drops from 3340 K to 3050 K. Note that the vertical dashed lines in Fig. 6.3 indicate the position of the SW front and highlight the zones of vibrational relaxation and dissociation (and electronic-translational relaxation).

Let us now consider how the refractive index in the relaxation zone behind the SW changes (see Fig. 6.3b). As well as for temperature profiles, on the $n-1$ profile both the vibrational relaxation and dissociation zones can be distinguished. This is quite understandable as the refractive index is linearly dependent on the gas density (see Eq. 4.1). Also, looking at Fig. 6.3b we can conclude that at the considered conditions the total refractivity is specified primarily by the $O_2(X^3\Sigma_g^-)$ contribution. So long as the dissociation degree is low, the contribution of $O(^3P)$ atoms is about 2.5% of the whole $n - 1$, and the aggregate contribution of excited electronic states of atomic and molecular oxygen is only 1.5%.

Since the contributions to $n - 1$ from the electronic states of O_2 and O_3 are not so substantial, we analyzed the influence of the dependence of the averaged polarizability on the vibrational T_{vib} and rotational T_{rot} temperatures only for the ground electronic state of molecular oxygen. To this end, the dashed curve in Fig. 6.3b shows the refractive index of the mixture obtained without allowance for the temperature dependence of the averaged polarizability for $O_2(X^3\Sigma_g^-)$: it was assumed here that the polarizability of $O_2(X^3\Sigma_g^-)$ rotational and vibrational states is equal to the polarizability of this molecule in the rovibrational state with $V = 0$ and $J = 0$. You can observe a slight difference in the calculated refractive indices just after the SW (with and without taking into account the temperature dependence of K^{GD}). This difference is governed by the dependence of the total K^{GD} on the rotational temperature ($T_{rot}(O_2(X^3\Sigma_g^-))$) immediately after the SW is equal to T). Then, with increasing distance from the SW front, this difference starts to grow as a result of vibrational-translational relaxation, reaches 1.5% (for $n - 1$) towards the end of the vibrational relaxation zone and practically remains constant afterward, only slightly falling (almost imperceptible to the eye) in consequence of temperature drop in the dissociation zone.

Thus, one can see that when interpreting refractivity measurements behind a SW front in oxygen, the allowance for each of the effects at issue (vibrational relaxation, dissociation, and electronic-translational relaxation) is essential. First of all, the change in $n - 1$ stems from the variation of the flow density during the relaxation of vibrational and electronic degrees of freedom of molecular gas species and their dissociation. Secondly, the fact that the dissociation products possess a larger polarizability than the source diatomic molecules (by 0.002 Å^3 per one $O_2(X^3\Sigma_g^-)$ molecule decay, see Sect. 6.1.2) also affects the refractivity variation. And, finally, the effect of vibrational and rotational temperatures on the K^{GD} constant plays a small (but not entirely negligible) role.

In principle, the latter could allow one the use of optical methods grounded on the refractivity to measure the population of vibrational states in molecular gases. Certainly, when analyzing the measurements obtained under similar post-SW conditions, it is difficult to achieve such accuracy of about $\sim 1\%$ for $n - 1$ to separate out the effect of nonequilibrium over the vibrational states on the K^{GD} magnitude. Perhaps, with more severe excitation of the components of the mixture, which can be realized under the conditions of laser or plasma exposure, in gas flows with intensive chemical reactions, and also behind ultra-strong SWs, the question of taking into account the complex dependence of the Gladstone–Dale constant on the thermal nonequilibrium effects may become more acute.

6.2   Intermolecular Potential

As is known [42, 43], based on the distance between the centers of mass of the interacting molecules, the intermolecular potential can be divided into distinctive regions. Forces acting at the short range are repulsive and result from the overlap of the molecular wave functions and symmetry requirements imposed by the Pauli exclusion principle. At longer range, electrostatic, induction, and dispersion attractive forces specified by the electric properties of colliding particles dominate [44, 45]. The elastic collision cross section for a pair of molecules, that, inter alia, govern the transport properties of molecular gases and the gas-kinetic collision frequency, depends substantially on the depth and specific shape of the long-range intermolecular potential.

The data on intermolecular potentials known in the literature often refer to molecules in the ground quantum states, but if the molecules are in excited states, the respective electric properties, by and large, differ from their ground-state counterparts, and the potential is to be accordingly modified. In general, the data on intermolecular potentials can be directly derived from molecular beam scattering experiments [43]. However, due to the difficulties in conducting appropriate experiments with excited molecules (resulting from their instability and from the impossibility to produce them in macroscopic quantities), theoretical determination of intermolecular potentials for excited molecules is particularly important for predictive modeling of gas-kinetic processes.

The authors are aware of some theoretical works in which the influence of internal degrees of freedom (vibrational [46–48] and rotational [46]) on the intermolecular potential was considered, however, for the most part (except for the work [48]), the dependence of electric properties on the level of excitation is not taken into account properly. With that, knowing the state-specific values of dipole moment and polarizability, it is not difficult to evaluate this effect in a simple manner.

As is widely assumed, for nonpolar neutral molecules, the intermolecular interaction is expressed by means of spherically symmetric Lennard-Jones (LJ) 12-6 potential [43, 49]. However for polar molecules, the orientation-dependent terms, responsible for dipole-induced dipole (polarization) and dipole-dipole (electrostatic) interactions must be involved as well [44]. In the general case, to describe the interaction of two arbitrary molecules (i and j), the dipole-reduced formalism method (DRFM) worked out by Paul and Warnatz [50] is to be invoked. In terms of DRFM, the Stockmayer 12-6-3 potential with the angle-dependent terms, once subjected to the thermally orientation-averaged procedure, can be reduced to the effective spherically symmetric LJ 12-6 potential (see [51] for details).

In compliance with this technique, the interaction between two polar (in the general case) molecules of i and j types at the distance r is determined by the effective spherically symmetrical LJ-type potential

$$\varphi_{ij}^{\text{eff}}(r) = C_6^{\text{eff}} \left(\frac{(\sigma_{ij}^{\text{eff}})^6}{r^{12}} - \frac{1}{r^6} \right) \tag{6.2}$$

with the following parameters

$$C_6^{\text{eff}} = C_6^{\text{disp}} \left(1 + \frac{C_6^{\text{ind}}}{C_6^{\text{disp}}} + \frac{C_6^{\text{el}}}{4\,C_6^{\text{disp}}} \right)^2, \tag{6.3}$$

$$\sigma_{ij}^{\text{eff}} = \sigma_{ij} \left(1 + \frac{C_6^{\text{ind}}}{C_6^{\text{disp}}} + \frac{C_6^{\text{el}}}{4\,C_6^{\text{disp}}} \right)^{-\frac{1}{6}}, \tag{6.4}$$

where σ_{ij} is the collision diameter of interacting particles, C_6^{disp}, C_6^{el}, and C_6^{ind} are the terms responsible for the dispersion interaction, electrostatic (dipole-dipole) interaction of dipoles, and polarization (dipole-induced dipole) interaction, respectively. These terms as consistent with [44, 51] can be defined as

$$C_6^{\text{disp}} = \frac{3}{2} E_h a_0^{3/2} \frac{\alpha_i \alpha_j}{\sqrt{\alpha_i/Z_i} + \sqrt{\alpha_j/Z_j}}, \tag{6.5}$$

$$C_6^{\text{el}} = \frac{2}{3\,k_b\,T} \mu_i^2 \mu_j^2, \tag{6.6}$$

$$C_6^{\text{ind}} = \mu_i^2 \alpha_j + \mu_j^2 \alpha_i. \tag{6.7}$$

Here E_h is the atomic energy unit (27.211 eV), a_0 is the Bohr (atomic) radius, α_i and α_j are the static isotropic polarizabilities, Z_i and Z_j are the numbers of valence electrons (the number of electrons in the outer sub-shell of molecule; see, for example, the classical book [52] for details), μ_i and μ_j are the dipole moments of particles of i and j sorts, k_b is the Boltzmann constant. In turn, the effective potential well depth for this case is stated as

$$\varepsilon_{ij} = \frac{C_6^{\text{eff}}}{4(\sigma_{ij}^{\text{eff}})^6}. \tag{6.8}$$

Notice that the Slater-Kirkwood expression [43, 53] is used here for C_6^{disp}, since the slight overestimation of dispersion term by the Slater-Kirkwood equation with respect to the well-known London formula can effectively compensate the neglecting the exchange forces in a long-range potential [51].

It is worth mentioning that for the ground state molecules the values of collision diameter σ_{ij} that specifies the range of repulsive valence forces in the Lennard-Jones potential can be obtained and tabulated via parameterization of experimental data on such macroscopic observables as second virial, viscosity, and diffusion coefficients [54] or results of specific quantum chemical calculations [55–57]. However, the problem of estimating the collision diameter of the vibrationally and rotationally excited species still has not a convincing solution [46, 58]. In [59] we employed the conjecture [58, 60], incidentally, shared by other researchers [61–63], that the gas

Fig. 6.4 The values of $r_{V,J}$ for H_2 calculated in [31]

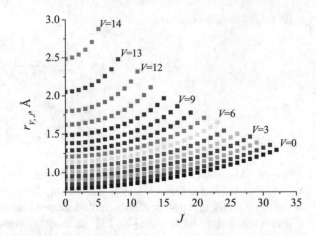

kinetic diameter of vibrationally and rotationally excited molecule is proportional to the averaged value of internuclear separation $r_{V,J}$ (r-centroids) for the given state specified by V and J quantum numbers.

It is well known that the $r_{V,J}$ quantity for diatomic molecules monotonically grows both with V (due to cubic anharmonicity of a potential and asymmetry about its minimum) and J increase (owing to centrifugal distortion) [64, 65]. As can be seen from Fig. 6.4 on the example of H_2 molecule, this growth can be very substantial. We also see there that the anharmonic vibrational effect is apparently more significant in terms of the influence on r_e than the centrifugal distortion.

The values of $r_{V,J}$ were determined in [31] for the different diatomic molecules considered in previous Chapters (see also Sect. 3.1.2). This enables us to evaluate the gas kinetic diameter for the molecules in the given rovibrational state. For this purpose, the state-specific collision diameter $\sigma_{ii}^{V,J}$ ($j = i$) can be calculated through the collision diameter of nonexcited molecule $\sigma_{ii}^{V=0, J=0}$ by

$$\sigma_{ii}^{V,J} = \sigma_{ii}^{V=0, J=0} + r_{V,J} - r_{V=0, J=0}. \tag{6.9}$$

The values of $\sigma_{ii}^{V=0, J=0}$ can be adopted from elsewhere [49, 66, 67] or, knowing the spatial structure of the molecule, they can be easily estimated by using the approach proposed by the authors [51, 68] and recently incorporated in the Chemcraft program [69] widely employed for handling the molecular structure.

Consider now the effective spherically symmetric LJ 12-6 potential

$$\varphi_{ij}^{LJ}(r) = 4\varepsilon_{ij} \left[\left(\frac{\sigma_{ij}^{eff}}{r} \right)^{12} - \left(\frac{\sigma_{ij}^{eff}}{r} \right)^{6} \right] \tag{6.10}$$

calculated utilizing the described methodology for some vibrationally excited diatomic molecules. Figure 6.5 depicts the effective $\varphi_{ii}^{LJ}(r)$ dependences for

Fig. 6.5 Effective $\varphi_{ii}^{LJ}(r)$ potentials for vibrationally excited H_2, OH, and HF molecules ($J = 0$, $T = 298$ K)

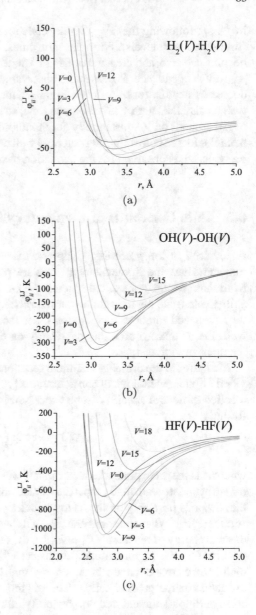

(a)

(b)

(c)

$H_2(V, J = 0)$, $OH(V, J = 0)$, and $HF(V, J = 0)$ molecules at $T = 298$ K. As seen, the positions of the minima of the intermolecular potential shift to the right with V growth, which is due to the monotonic increase in the equilibrium interatomic distance $r_{V,J}$ (owing to cubic anharmonicity). At the same time, for the depth of the $\varphi_{ii}^{LJ}(r)$ well, the dependence on the number of vibrational level can be more complex.

In particular, for nonpolar molecules, such as H_2, the depth of the potential well ε_{ii} is due to only the dispersion term governed by Eq. (6.5), therefore, with an increase in V, ε_{ii}, following the $\alpha_{J=0}^{\mathrm{PP}}(V)$ dependence (see Fig. 3.10), increases as a whole (till not too high levels). For polar molecules, the importance of C_6^{el} and C_6^{ind} terms become decisive, and the nature of $\varepsilon_{ii}(V)$ dependence is primarily determined by the $\mu_{J=0}^0(V)$ dependence. So, since the observable dipole moment for the OH molecule decreases steadily for $V > 1$ (see Fig. 3.4), the depth of the corresponding potential well ε_{ii} also falls with V. At the same time, so long as the observable dipole moment of HF depends on V substantially nonmonotonically, we observe in Fig. 6.5c an alike behavior for the corresponding potential well depth. Something similar, although less pronounced, is observed for the excitation of rotational degrees of freedom as well.

6.3   Rate Constants of Elementary Reactions

As evinced in the preceding Chapters, the vibrationally and rotationally excited molecules have the dipole moment and static polarizability that is, in general, distinct from those for the nonexcited species. As far as dipole moment and dipole polarizability determine the intermolecular interaction [44], the excitation of the molecular vibrations and rotations can, among other factors, affect the chemical reactivity as well. Let us discuss now this possibility on the example of bimolecular chemical reactions of various types.

First, turn to the ordinary chemical reactions between neutral molecules. As provided by phenomenological considerations [70], the rate constant of the bimolecular reaction of neutral species can be expressed by virtue of the established Arrhenius formula

$$k_{\mathrm{neut}}(T) = A \exp\left(-\frac{E_a}{k_b\,T}\right),$$

where A is the preexponential factor that is weakly temperature dependent and determined by the structure and properties of reactants, whereas E_a is the activation energy that depends primarily on the energy barrier on the corresponding potential energy surface (PES). When the energy barrier of a chemical reaction is large compared to the thermal energy of reactants ($E_a > k_b\,T$), the reaction rate constant is predominately governed by the E_a magnitude. In this case, the excitation of molecular vibrations mainly leads to a decrease in the energy threshold and the impact of the molecular excitation on the preexponential factor is minor. For this reason, the effect of alteration of dipole moment and polarizability of reacting molecules on A as a result of excitation of internal degrees of freedom is generally ignored even in nonequilibrium chemical kinetics [70–72]. We note, in passing, that the quantitative description of the influence of exciting the vibrational and rotational states of reacting molecules on the rate of chemical reactions with a nonzero energy barrier has been debated for many years, and several models intended to describe this effect were suggested by different research groups (see, for example [72–75]).

However, many significant elementary reactions proceed without a substantial energy barrier. For them, it is the long-range interaction between reacting particles that is determining factor for the reaction rate constant [76, 77]. The rate constant of a barrier-free reaction for neutral species can most easily be estimated by means of the capture approximation [76, 78]. In doing so, the pairwise interaction of polar (in the general case) particles can be characterized by the effective spherically symmetric potential involving the terms responsible for dispersion, dipole-induced dipole, and dipole-dipole interactions.

In compliance with this approximation, the capture happens when the reacting system moves inside a sphere with some critical radius of δ representing the distance between molecules, and the value of δ is related to the location of the effective centrifugal barrier for this process. Upon capture, the formation of product molecules is supposed to occur with the efficiency close to unity. As provided by the capture approximation [79, 80], the temperature-dependent rate constant $k_{neut}(T)$ of bimolecular barrier-free reaction of two neutral reactants that interact by means of the potential given by Eq. (6.2), can be stated as follows:

$$k_{neut}(T) = 1.706 \frac{Q_e^{PES}}{Q_e^i Q_e^j} \left(\frac{(C_6^{eff})^2 k_b T}{M^3} \right)^{1/6}. \tag{6.11}$$

Here M is the reduced molecular mass of reacting particles, C_6^{eff} is the effective parameter defined by Eq. (6.3), Q_e^i, Q_e^j and Q_e^{PES} are the electronic partition functions of the ith and the jth reactants and the electronic weight of reactive PES, respectively.

As shown above, the excitation of vibrational and rotational states of molecules can alter the averaged polarizability and dipole moment, and, hence, affect the rate coefficients of a chemical reaction. Dimensional analysis of Eqs. (6.2)–(6.8), and (6.11) indicates that, for the barrierless process, the rate constant is more sensitive to the variation of the product of dipole moments of reacting molecules ($k_{neut}(T) \sim (\mu_i \mu_j)^{2/3}$) rather than to the variation of the product of their polarizabilities ($k_{neut}(T) \sim (\alpha_i \alpha_j)^{1/3}$). Consequently, the effect of molecular excitation on the k_{neut} rate coefficient is to be more dramatic if both reactants have permanent dipole moments.

Let us examine now the influence of the excitation of OH radical vibrations on the rate constant for the following barrier-free reaction

$$OH(V) + HO_2 = H_2O + O, \tag{R1}$$

essential for atmospheric and combustion chemistry. Figure 6.6 represents the temperature-dependent R1 reaction rate constant calculated in compliance with Eq. (6.11) for $V = 0$, 10, and 15. The experimental evidence for the thermal rate constant of this reaction [81–83] is also given there. We see that the considered methodology reasonably reproduces the measurements, especially in view of the fact that there is no certainty that in these experiments [81–83] a reaction rate constant precisely with the thermalized OH molecules was measured. At relatively low temperatures ($T < 500$ K), when the contribution of the dipole-dipole term is particularly important (see Eq. 6.6), the excitation of OH vibrations ($V = 10$ and 15)

Fig. 6.6 Predicted state-specific constants for the reaction OH(V) + HO$_2$ = H$_2$O + O at $V = 0$, 10, and 15 as a function of gas temperature (curves) and the experimental data for the thermal equilibrium rate constant [81–83] (symbols) (Reused with permission from Ref. [59]. Copyright 2016 IOP Publishing Ltd)

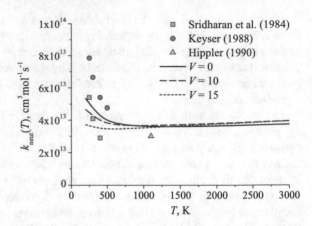

leads to a decrease in the averaged dipole moment of OH (see Chap. 3), that brings about a visible reduction of the reaction rate coefficient (by 0.7 times at $V = 15$). At higher temperatures ($T > 500$ K), when the dipole-dipole contribution vanishes, the effect of vibrational excitation becomes much weaker. The latter, in this case, is solely due to the increase in the observable polarizability of vibrationally excited OH(V) with V increment. Thereby, it was shown that the vibrational excitation of neutral molecules can slightly influence the rate constants of barrier-free chemical reactions between them.

Now we turn to the effect of molecular excitation on the rate constant of binary ion-molecule reactions, which are relevant for low-temperature plasma. The rate constant of binary ion-molecule reactions, in the case of low-energetic collisions (up to 1 eV), can be derived within the framework of statistical theory [84, 85]. For exothermic ion-molecule process, the reaction rate constant can be found by virtue of the Langevin capture model [70, 86] given by

$$k_{\text{ion-mol}}(T) = 2\pi q\left(\frac{\alpha_{\text{N}}}{M}\right)^{1/2} f(x)\frac{Q_e^{\text{PES}}}{Q_e^{\text{I}}Q_e^{\text{N}}}, \quad x = \frac{\mu_{\text{N}}}{(2\alpha_{\text{N}}k_b T)^{1/2}}, \qquad (6.12)$$

where q is the charge of ion, α_{N} and μ_{N} are the polarizability and dipole moment of neutral molecule, M is the reduced mass of colliding particles, $f(x)$ is the polarity factor responsible for orientation-dependent interaction, x is a dimensionless parameter, Q_e^{I} and Q_e^{N} are the electronic partition functions of the ion and neutral molecule. The polarity function $f(x)$ tabulated in [87] can be written for a practically interesting range of x values ($0 \leq x \leq 9$) in the following polynomial form [59]: $f(x) = 0.9775 + 0.1479x + 0.0863x^2 - 0.0054x^3$. However, for a broad range of x, we recommend using the following approximation based on tabulated values [87]:

$$f(x) = 1 + 0.6058\frac{x^2}{x + 2.1764}. \qquad (6.13)$$

Fig. 6.7 Rate constant for the reaction HCl(V)+C$^+$ → products at $V = 0$ and 15 as a function of T. The measured thermal rate constant [88, 89] is depicted by symbols

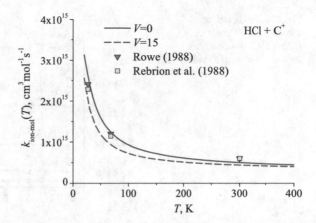

When analyzing Eqs. (6.11) and (6.12), we are led to infer that the rate constant of the ion-molecule reaction is more susceptible to the deviations of the polarizability of a neutral vibrationally excited reactant (in this case, we have $k_{\text{ion}-\text{mol}}(T) \sim \alpha_N^{1/2}$) than that of the reaction between neutral species ($k_{\text{neut}}(T) \sim \alpha_i^{1/3}$). In addition, it should be stressed that at the low-temperature limit or, equivalently, for the strongly polar molecules (i.e. for $x \gg 1$), the $k_{\text{ion}-\text{mol}}$ dependence on polarizability of a neutral reactant disappears, and we have the following asymptotics: $k_{\text{ion}-\text{mol}}(T) \sim \mu_N$ (whereas $k_{\text{neut}}(T) \sim \mu_i^{2/3}$). Hence, for this class of elementary processes, the influence of the excitation of molecular vibrations on the rate constant must be more intense compared to that for the reactions of neutral molecules.

Shown in Fig. 6.7 is the temperature-dependent rate constant of the ion-molecule reaction

$$\text{HCl}(V) + \text{C}^+ \rightarrow \text{ products} \tag{R2}$$

at room temperature and lower predicted employing the described methodology in [59] and measured elsewhere [88, 89] for $V = 0$. Note also that this reaction is of importance for interstellar and circumstellar gas-phase chemistry. We see sound agreement between the predictions and the measurements, which indicates the validity of using the described methodology.

Figure 6.7 also shows the $k_{\text{ion}-\text{mol}}(T)$ dependence of this reaction calculated for the case of strong vibrational excitation of HCl(V) ($V = 15$). We see that the rate constant for the interaction of HCl($V = 15$) with C$^+$ ion is appreciably smaller than that of the HCl($V = 0$) one, particularly lower temperatures. This is due to a diminishing of the averaged dipole moment of the HCl molecule with an increase in V and, accordingly, decrease in $f(x)$ factor in Eq. (6.12). In addition, Fig. 6.8 represents the dependence of this rate constant on V for various translational temperatures. We can observe that the excitation of HCl vibrations can lead to the sizeable variation of the ion-molecule reaction rate constant just at very low temperatures.

Fig. 6.8 Predicted rate constant for the reaction $HCl(V)+C^+ \to$ products versus V for fixed T values

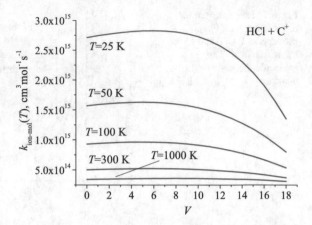

Fig. 6.9 Predicted rate constant for the reaction $CO(V) + H_3^+ \to$ products versus V for fixed T values. The experimental [92, 93] and accurate theoretical [91] values for thermal rate constant at $T = 300\,K$ are depicted by symbols

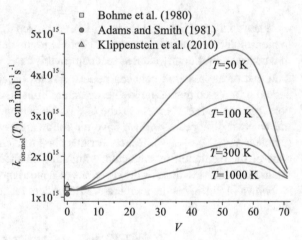

Another ion-molecule reaction that would be interesting to consider in the sense of the influence of vibrational excitation of reacting neutral molecule is the reaction of a CO molecule with H_3^+ ion. This process is also highly important for interstellar chemistry since it is responsible (alongside $O(^3P)+H_3^+$ process) for the destruction of H_3^+ in dense interstellar clouds [90, 91]. As the absolute value of the averaged dipole moment of CO varies broadly (0-0.99 D) depending on V [31] (see Chap. 3), the dependence of this reaction rate constant on V can be sharp. It is visualized in Fig. 6.9 for different values of gas temperature. The measurements for the reaction of H_3^+ positive ion with a nonexcited CO molecule [92, 93] together with the accurate predictions of Klippenstein et al. [91] are provided there as well. We observe that a stronger dependence of the rate constant on V corresponds to a lower gas temperature. Specifically, at $T = 50$ K, the excitation of the CO molecule to the 20th vibrational level leads to a two-fold increase in the reaction rate constant.

Thus, the effect of the excitation of diatomic molecules for ion-molecule reactions can be more pronounced than that for the reactions involving only neutral particles, especially at very low temperatures, where the influence of dipole-dipole interactions is most essential.

As concerns the reaction kinetics with electronically excited species, this is a particular and challenging area of chemical physics, which is not considered in this book. Even though there are analytic models for calculating the rate constants of chemical reactions involving electronically excited molecules, for instance, the Modified Model of Vibronic Terms [94], even the modest authors' experience exhibits that without precision experiments or special comprehensive quantum chemical studies it is impossible to obtain even qualitative estimates of the effect of electronic excitation of components on their reactivity [5, 30, 95–98].

6.4   Transport Properties

Calculation of transport properties of molecular gases under nonequilibrium conditions, when the vibrational (or even rotational) temperature deviates from translational one, is highly needed, for instance, for chemical kinetic modeling of the processes behind SWs and in the supersonic expanding jets [99–101]. The point is that if the molecules are in excited states, their electric properties and, accordingly, intermolecular potentials (see Sect. 6.2) differ from their ground-state counterparts. Hence, the excitation of molecules can lead to a change in their transport coefficients.

Previously, some theoretical studies were focused on estimating the transport properties of vibrationally and rotationally excited molecules [46, 58, 60–62, 102], but in these studies, the dependence of electric properties on the level of excitation was not taken into account properly (in fact, only the influence of molecular vibrations on the collision diameter was allowed for). At the same time, the modification of electric properties of molecules owing to excitation to the high vibrational and rotational states can alter the intermolecular potential, and, consequently, the collision integrals. This effect undoubtedly should be also accounted for when evaluating the influence of the excitation of internal degrees of freedom of molecules on their transport coefficients.

Recently, the path to solving the problem was outlined in [59]. Let us briefly consider the methodology proposed there. The standard kinetic theory of gases makes it possible to evaluate the transport properties of gaseous species granted the potential of interparticle interaction is established. Particularly, binary diffusion coefficient D_{ij} can be estimated with the help of a standard equation obtained based on the Chapman-Enskog theory [42, 103]. The expression suitable for low and moderate pressures is as follows [42]:

$$D_{ij} = \frac{3}{8} \frac{\sqrt{2\pi M_{ij}\, k_b\, T / N_A}}{\pi \sigma_{ij}^2} \frac{1}{\Omega^{(1,1)*}\left(\varepsilon_{ij}, T\right)} \frac{RT}{P_0 2 M_{ij}}. \tag{6.14}$$

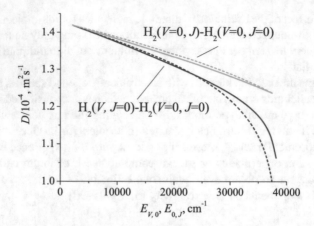

Fig. 6.10 Dependence of binary diffusion coefficients on the energy of vibrational $E_{V,J=0}$ or rotational $E_{V=0,J}$ levels for the $H_2(V,J = 0)$-$H_2(V = 0, J = 0)$ and $H_2(V = 0, J)$-$H_2(V = 0, J = 0)$ molecular pairs at $T = 300$ K (solid curves). Diffusion coefficients obtained under the assumption of fixed $\Omega^{(1,1)*}$ values are depicted by dotted curves (Adapted with permission from Ref. [59]. Copyright 2016 IOP Publishing Ltd)

Here M_{ij} is the reduced molar mass of colliding particles of i and j sort, σ_{ij} is their collision diameter, P_0 is the normal pressure, R is the universal gas constant, $\Omega^{(1,1)*}(\varepsilon_{ij}, T)$ is the reduced collision integral normalized by the cross sections for rigid spheres of diameter σ_{ij}. In fact, the product $\sigma_{ij}^2 \, \Omega^{(1,1)*}(\varepsilon_{ij}, T)$ is the energy-averaged cross section that depends on the temperature and interparticle potential φ_{ij}. If this potential can be described in the form of the effective spherically symmetrical LJ-type potential given by Eq. (6.2), the values of $\Omega^{(1,1)*}(\varepsilon_{ij}, T)$ dependent on the effective potential well depth (see Eq. 6.8) can be adopted from [104].

Of most general interest are the binary diffusion coefficients for molecules excited to the specific vibrational-rotational state with V and J numbers $AB(V, J)$ in the bath gas of the ground state molecules $AB(0, 0)$. Displayed in Fig. 6.10 are the binary diffusion coefficients for the $H_2(V, 0)$-$H_2(0, 0)$ and $H_2(0, J)$-$H_2(0, 0)$ molecular pairs against the energy of vibrational $E_{V,0}$ and rotational $E_{0,J}$ levels calculated considering the effect of molecular excitation on both the collision diameter and the collision integral. For comparison, the same figure also shows the diffusion coefficients obtained assuming that the collision integral is fixed. We may see that the excitation of H_2 molecule to upper rotational and vibrational states results in the marked decrease of the binary diffusion coefficient, however the role of the modification of the collision integral in this case is minimal. Thereby, the influence of the excitation of H_2 on collision integral is small, and the change in binary diffusion coefficient for this nonpolar compound upon excitation of vibrations and rotations is primarily specified by an increase in its effective gas kinetic cross section. Notice that the variation in the diffusion coefficient as a result of the vibrational excitation of H_2 is more marked than upon excitation of rotations (cf. Fig. 6.4).

Fig. 6.11 Dependence of binary diffusion coefficients for the HCl($V, J = 0$)-HCl($V = 0$, $J = 0$)/O$_2$($V = 0$, $J = 0$) (**a**) and HF($V, J = 0$)-HF($V = 0$, $J = 0$)/O$_2$($V = 0$, $J = 0$) (**b**) pairs of molecules on the V number at $T = 300$ K (solid curves). Diffusion coefficients calculated under the assumption of fixed $\Omega^{(1,1)*}$ values are depicted by dotted curves (Adapted with permission from Ref. [59]. Copyright 2016 IOP Publishing Ltd)

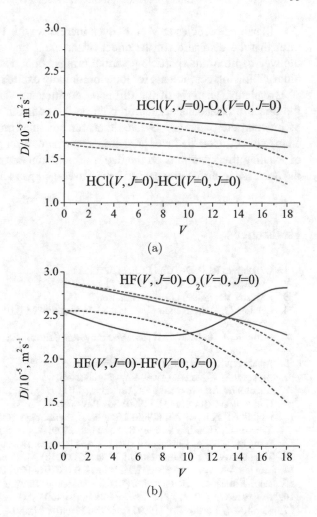

(a)

(b)

We next discuss a gas made up of polar molecules. Figure 6.11 shows the binary diffusion coefficients of vibrationally excited but rotationally cold molecules HCl(V, $J = 0$) and HF(V, $J = 0$) in a diluent gas of nonexcited molecules (HCl, HF, and O$_2$) against V at room temperature. As described in Chap. 3, the dipole moment of these diatomic molecules, after averaging over the vibrational motion, exhibits a sharp dependence on V. Consequently, intense dipole-dipole and dipole-induced dipole interactions affect the collision integral $\Omega^{(1,1)*}$ appreciably, and this brings about a change in diffusion coefficient with V growth.

The plots provided in Fig. 6.11 suggest that the effect of vibrational excitation of HCl molecules on their electric properties virtually compensates for the decrease in the binary diffusion coefficient through the increment of the averaged collision diameter. Upon the vibrational excitation of the HF molecule, having a higher dipole moment than the HCl one, the diffusion coefficient for the HF($V, J = 0$)-HF($V = 0$,

$J = 0$) pair of particles at $V > 10$ even increases with V number. Thereby, we can infer that the disregard for the effect of the excitation of molecular vibrations on the well depth of interparticle potential brings about a significant underestimate of binary diffusion coefficients for the vibrationally excited polar diatomic molecules.

Eventually, analysis of the diffusion coefficients variation with allowance for the effect of molecular excitation both on the collision diameter and on the depth of interparticle potential, revealed that, for nonpolar molecules, the impact of the change of collision diameter predominates, whereas, for polar species, the influence of exciting the vibrations on the depth of potential can counteract or even surpass the decrease in diffusivity caused by the averaged gas-kinetic diameter growth.

References

1. Cvetanovic RJ (1974) Can J Chem 52:1452
2. Osipov AI, Uvarov AV (1992) Sov Phys Usp 35:903
3. Azyazov VN (2009) Quantum Electron 39:989
4. Colonna G, D'Ammando G, Pietanza LD, Capitelli M (2015) Plasma Phys Control Fusion 57: 14009
5. Starik AM, Loukhovitski BI, Sharipov AS, Titova NS (2015) Phil Trans R Soc A 373:20140341
6. Alpher RA, White DR (1959) Phys Fluids 2:153
7. Alpher RA, White DR (1959) Phys Fluids 2:162
8. Kharitonov AI, Khoroshko KS, Shkadova VP (1974) Fluid Dyn 9:851
9. Gladkov SM, Koroteev NI (1990) Sov Phys Usp 33:554
10. Osipov AI, Filippov AA (1989) J Eng Phys Thermophys 56:590
11. Yun-yun C, Zhen-hua L, Yang S, An-zhi H (2009) Appl Opt 48:2485
12. Wang M, Mani A, Gordeyev S (2012) Annu Rev Fluid Mech 44:299
13. Tropina AA, Wu Y, Limbach CM, Miles RB (2018) AIAA paper, pp 3904
14. Tropina AA, Wu Y, Limbach CM, Miles RB (2019) J Phys D: Appl Phys 53:105201
15. Takahashi Y, Yamada K, Abe T (2014) J Spacecraft Rockets 51:1954
16. Askaryan GA (1966) JETP Lett (in Russian) 4:400
17. Starik AM, Taranov OV (1997) Quantum Electron 27:550
18. Bader RFW, Keith TA, Gough KM, Laidig KE (1992) Mol Phys 75:1167
19. Buckingham AD (1956) Trans Faraday Soc 52:1035
20. Hohm U (1994) J Chem Phys 101:6362
21. Chattaraj PK, Poddar A (1999) J Phys Chem A 103:1274
22. Jaque P, Toro-Labbe A (2002) J Chem Phys 117:3208
23. Chattaraj PK, Roy DR, Elango M, Subramanian V (2005) J Phys Chem A 109:9590
24. Krech RH, McFadden DL (1977) J Am Chem Soc 99:8402
25. Sabirov DS, Garipova RR, Cataldo F (2018) Mol Astrophys 12:10
26. Ruscic B, Bross DH (2016) Active thermochemical tables (ATcT) values based on ver. 1.122 of the thermochemical network. https://atct.anl.gov/
27. Lee JS (2005) Theor Chem Acc 113:87
28. Schwerdtfeger P, Nagle JK (2019) Mol Phys 117:1200
29. Hohm U (2013) J Mol Struct 1054–1055:282
30. Pelevkin AV, Sharipov AS (2018) J Phys D: Appl Phys 51:184003
31. Loukhovitski BI, Sharipov AS, Starik AM (2016) J Phys B: At Mol Opt Phys 49:125102
32. Hohm U (2000) J Phys Chem A 104:8418
33. Blair SA, Thakkar AJ (2013) Chem Phys Lett 556:346

34. Huber KP, Herzberg G (1979) Molecular spectra and molecular structure, vol 4. Constants of diatomic molecules. Van Nostrand Reinhold, New York
35. Sharipov AS, Loukhovitski BI, Pelevkin AV, Kobtsev VD, Kozlov DN (2019) J Phys B: At Mol Opt Phys 52:045101
36. Kadochnikov IN, Arsentiev IV (2018) J Phys D: Appl Phys 51:374001 (22 pp)
37. Kadochnikov IN, Loukhovitski BI, Starik AM (2015) Plasma Sources Sci Technol 24:055008 (14 pp)
38. Starik AM, Titova NS (2001) Tech Phys 46(8):929
39. Miller TM (2010) CRC handbook of chemistry and physics, 90th edn., vol. 10, chap. Atomic and molecular polarizabilities. CRC Press, Boca Raton, Florida, pp 193–202
40. Andersson K, Sadlej AJ (1992) Phys Rev A 46:2356
41. Sharipov AS, Loukhovitski BI, Starik AM (2017) J Phys B: At Mol Opt Phys 50:165101(19pp)
42. Hirschfelder JO, Curtiss CF, Bird RB (1954) Molecular theory of gases and liquids. Wiley NY; Chapman and Hall, London
43. Buckingham AD, Fowler PW, Hutson JM (1988) Chem Rev 88:963
44. Kaplan IG (2006) Intermolecular interactions: physical picture, computational methods and model potentials. Wiley, Hoboken, NJ
45. Gould T, Bucko T (2016) J Chem Theory Comput 12:3603
46. Kunc JA (1990) J Phys B: At Mol Opt Phys 23:2553
47. Damyanova M, Zarkova L, Hohm U (2009) Int J Thermophys 30:1165
48. Bykov AD, Klimeshina TE, Rodimova OB (2014) 20th international symposium on atmospheric and ocean optics: atmospheric physics. International Society for Optics and Photonics, pp 92,920P–92,920P–8
49. Brown NJ, Bastien LAJ, Price PN (2011) Prog Energy Combust Sci 37:565
50. Paul P, Warnatz J (1998) Proc Combust Inst 27:495
51. Sharipov AS, Loukhovitski BI, Tsai CJ, Starik AM (2014) Eur Phys J D 68(4):99
52. Hellmann HGA (1944) Einführung in die Quantenchemie. Michigan, Ann Arbor
53. Salem L (1960) Mol Phys 3:441
54. Bzowski J, Kestin J, Mason EA, Uribe FJ (1990) J Phys Chem Ref Data 19:1179
55. Shadman M, Yeganegi S, Ziaic F (2009) Chem Phys Lett 467:237
56. Jasper AW, Miller JA (2014) Combust Flame 161:101
57. Mehio N, Dai S, Jiang D (2014) J Phys Chem A 118:1150
58. Kang SH, Kunc JA (1991) J Phys Chem 95:6971
59. Sharipov AS, Loukhovitski BI, Starik AM (2016) J Phys B: At Mol Opt Phys 49:125103
60. Gorbachev YE, Gordillo-Vaszquez FJ, Kunc JA (1997) Physica A 247:108
61. Galkin VS, Makashev NK, Rastigecv EA (1996) Fluid Dyn 31:144
62. Kustova EV, Kremer GM (2015) Chem Phys Lett 636:84
63. Kremer GM, Kunova OV, Kustova EV, Oblapenko GP (2018) Phys A 490:92
64. Toyama M, Oka T, Morino Y (1964) J Mol Spectosc 13:193
65. Lounila J, Wasser R, Diehl P (1987) Mol Phys 62:19
66. Svehla RA (1962) Estimated viscosities and thermal conductivities of gases at high temperatures. Technical Report, NASA Technical Report R-132
67. Kee RJ, Dixon-Lewis G, Warnatz J, Coltrin ME, Miller JA, Moffat HK (1988) A fortran computer code package for the evaluation of gas-phase, multicomponent transport properties. Technical Report. Sandia National Laboratories, Livermore, CA
68. Sharipov AS, Loukhovitski BI (2019) Struct Chem 30:2057
69. Andrienko GA, Chemcraft version 1.8. http://www.chemcraftprog.com
70. Fernandez-Ramos A, Miller JA, Klippenstein SJ, Truhlar DG (2006) Chem Rev 106:4518
71. Capitelli M, Ferreira CM, Gordiets BF, Osipov AI (2000) Plasma kinetics in atmospheric gases, Springer series on atomic, optical, and plasma physics, vol 31. Springer, Berlin.
72. Fridman A (2008) Plasma chemistry. Cambridge University Press, Cambridge, UK
73. Lifshitz A, Teitelbaum H (1997) Chem Phys 219(2–3):243
74. Da Silva ML, Guerra V, Loureiro J (2007) Chem Phys 342:275
75. Arsentiev IV, Loukhovitski BI, Starik AM (2012) Chem Phys 398:73

76. Clary DC (1984) Mol Phys 53:3
77. Sumathi R, Green WH Jr (2002) Theor Chem Acc 108:187
78. Faure A, Wiesenfeld L, Valiron P (2000) Chem Phys 254:49
79. Yang K, Ree T (1961) J Chem Phys 35:588
80. Sharipov AS, Starik AM (2015) J Phys Chem A 119:3897
81. Sridharan UC, Qiu LX, Kaufman F (1984) J Phys Chem 88:1281
82. Keyser LF (1988) J Phys Chem 92:1193
83. Hippler H, Troe J, Willner J, Ber (1990) Bunsenges Phys Chem 93:1755
84. Smirnov BM (1974) Ions and excited atoms in a plasma. Atomizdat, Moscow (in Russian)
85. Bates DR (1978) Proc R Soc Lond A 360:1
86. Su T (1988) J Chem Phys 88:4102
87. Chernyi GG, Losev SA, Macheret SO, Potapkin BV (2002) Prog Astronaut Aeronaut. Reston, V.A: AIAA 196, 311
88. Rowe BR (1988) Astrophysics and space science library. V. 146. Rate coefficients in astrochemistry. Studies of ion-molecule reactions at $T < 80$ K. Springer Netherlands, Dordrecht
89. Rebrion C, Marquette JB, Rowe BR (1988) Chem Phys Lett 143:130
90. Petrie S, Bohme DK (2007) Mass Spectrom Rev 26:258
91. Klippenstein SJ, Georgievskii Y, McCall BJ (2010) J Phys Chem A 114:278
92. Bohme DK, Mackay GI, Schiff HI (1980) J Chem Phys 73:4976
93. Adams NG, Smith D (1981) Astrophys J 248:373
94. Starik A, Sharipov A (2011) Phys Chem Chem Phys 13:16424
95. Sharipov A, Starik A (2011) J Phys Chem A 115(10):1795
96. Sharipov AS, Starik AM (2012) J Phys Chem A 116:8444
97. Sharipov AS, Loukhovitski BI, Starik AM (2016) J Phys Chem A 120:4349
98. Pelevkin AV, Loukhovitski BI, Sharipov AS (2017) J Phys Chem A 121:9599
99. Adamovich IV, Macheret SO, Rich JW (1994) Chem Phys 182:167
100. Armenise I, Barbato M, Capitelli M, Kustova E (2006) J Thermophys Heat Transfer 20:465
101. Capitelli M, Armenise I, Bisceglie E, Bruno D, Celiberto R, Colonna G, D'Ammando G, Pascale OD, Esposito F, Gorse C, Laporta V, Laricchiuta A (2012) Plasma Chem Plasma Process 32:427
102. Kustova EV (2001) Chem Phys 270:177
103. Capitelli M, Laricchiuta A, Bruno D (2013) Fundamental aspects of plasma chemical physics: transport. Springer
104. Neufeld PD, Janzen AR, Aziz RA (1972) J Chem Phys 57:1100

Chapter 7
Conclusions and Future Prospects

In this book, we have seen that the molecular electrical properties, namely, the dipole moment and polarizability, substantially depend on the populations of the rotational, vibrational, and electronic quantum states of a given molecule. Thus, some "observable" properties of molecules determined precisely by their electric response characteristics, such as the refractive index, dielectric constant, transport coefficients, and rate constants of chemical reactions, can largely depend on the specific distribution of molecules over rotational, vibrational, and electronic levels. In this connection, of particular interest are the essentially thermal nonequilibrium distributions, which are realized under specific conditions, relevant to the strong shock waves, rapidly expanding gas jets, upper and middle atmospheres of different planets, electric discharges, gas lasers, etc.

Therefore, accurate knowledge of state-specific electric properties on the one hand, and the construction of accurate models of thermally nonequilibrium kinetics of the predictive level on the other, make it possible to comprehensively take into account the effect of internal degrees of freedom on various properties of molecular gases (optic, electric, transport, reactivity) in any given conditions.

In this summary, we have tried to resume and discuss the existing knowledge on these issues with an emphasis on the authors' own experience [1–5]. The focus of the present contribution is on methodological and computational aspects of accurate theoretical determination of different state-specific molecular electric response properties by means of quantum chemistry methods.

Lastly, we can distinguish several topical areas of further development of this field. First, a consistent account of the effect of rotational motion and an accurate determination of the electric properties for high-lying vibrational levels of polyatomic molecules are needed. Secondly, of considerable interest is the determination of the state-specific values of the dynamic part of the polarizability, which is important for the interpretation of optical measurements. Thirdly, it is advisable to continue

© The Author(s), under exclusive license to Springer Nature Switzerland AG 2022   97
A. S. Sharipov et al., *Influence of Internal Degrees of Freedom on Electric and Related Molecular Properties*, SpringerBriefs in Electrical and Magnetic Properties of Atoms, Molecules, and Clusters, https://doi.org/10.1007/978-3-030-84632-9_7

studying the effect of electronic excitation on electrical properties. Finally, the study of the effect of ionization of molecules and atoms on the refractivity and other properties of nonequilibrium air plasma is also of great practical importance.

We hope that by listing the possible directions for further research, we will encourage theoreticians and, importantly, experimentalists to further investigate the effect of internal degrees of freedom of molecules and their ions on the electric and related properties of gases.

References

1. Loukhovitski BI, Sharipov AS, Starik AM (2016) J Phys B: At Mol Opt Phys 49:125102
2. Sharipov AS, Loukhovitski BI, Starik AM (2016) J Phys B: At Mol Opt Phys 49:125103
3. Sharipov AS, Loukhovitski BI, Starik AM (2017) J Phys B: At Mol Opt Phys 50:165101(19pp)
4. Sharipov AS, Loukhovitski BI, Pelevkin AV, Kobtsev VD, Kozlov DN (2019) J Phys B: At Mol Opt Phys 52:045101
5. Lukhovitskii BI, Sharipov AS, Arsent'ev IV, Kuzmitskii VV, Penyazkov OG (2020) J Eng Phys Thermophys 93:850

Printed in the United States
by Baker & Taylor Publisher Services